7-11

we
:er

Outs

DATE DUE

APR 2 5 2012		
FEB 1 1 2013		
OCT 0 8 2013		
DEC 0 3 2014		
NOV 0 8 2016		

A PERIGEE BOOK

A PERIGEE BOOK
Published by the Penguin Group
Penguin Group (USA) Inc.
375 Hudson Street, New York, New York 10014, USA
Penguin Group (Canada), 90 Eglinton Avenue East, Suite 700, Toronto, Ontario M4P 2Y3, Canada
(a division of Pearson Penguin Canada Inc.)
Penguin Books Ltd., 80 Strand, London WC2R 0RL, England
Penguin Group Ireland, 25 St. Stephen's Green, Dublin 2, Ireland (a division of Penguin Books Ltd.)
Penguin Group (Australia), 250 Camberwell Road, Camberwell, Victoria 3124, Australia
(a division of Pearson Australia Group Pty. Ltd.)
Penguin Books India Pvt. Ltd., 11 Community Centre, Panchsheel Park, New Delhi—110 017, India
Penguin Group (NZ), 67 Apollo Drive, Rosedale, North Shore 0632, New Zealand
(a division of Pearson New Zealand Ltd.)
Penguin Books (South Africa) (Pty.) Ltd., 24 Sturdee Avenue, Rosebank, Johannesburg 2196, South Africa
Penguin Books Ltd., Registered Offices: 80 Strand, London WC2R 0RL, England

While the author has made every effort to provide accurate telephone numbers and Internet addresses at the time of publication, neither the publisher nor the author assumes any responsibility for errors, or for changes that occur after publication. Further, the publisher does not have any control over and does not assume any responsibility for author or third-party websites or their content.

PRINTING HISTORY
Perigee hardcover edition / April 2010
Perigee trade paperback edition / January 2011

Perigee trade paperback ISBN: 978-0-399-53638-0

The Library of Congress has cataloged the Perigee hardcover edition as follows:

Colby, Brandon.
   Outsmart your genes : how understanding your DNA will empower you to protect yourself against cancer, Alzheimer's, heart disease, obesity, and many other conditions / Brandon Colby.— 1st ed.
      p.   cm.
   "A Perigee book."
   Includes bibliographical references and index.
   ISBN 978-0-399-53557-4
   1. Genetics—Popular works. 2. DNA—Popular works. I. Title.
   QH437.C65 2010
   616'.042—dc22   2009050863

PRINTED IN THE UNITED STATES OF AMERICA

10 9 8 7 6 5 4 3 2 1

Neither the publisher nor the author is engaged in rendering professional advice or services to the individual reader. The ideas, procedures, and suggestions contained in this book are not intended as a substitute for consulting with your physician. All matters regarding your health require medical supervision. Neither the author nor the publisher shall be liable or responsible for any loss or damage allegedly arising from any information or suggestion in this book.

Most Perigee books are available at special quantity discounts for bulk purchases for sales promotions, premiums, fund-raising, or educational use. Special books, or book excerpts, can also be created to fit specific needs. For details, write: Special Markets, Penguin Group (USA) Inc., 375 Hudson Street, New York, New York 10014.

To my mother and father—
Thank you for your limitless love,
for nurturing my curiosity,
and for teaching me that we all
have the ability to outsmart our genes.

Acknowledgments

My heartfelt thanks go out to a number of people without whom I wouldn't have been able to put mind to matter, pen to paper, and write. First and foremost, thank you to my family, including those related to me by the bond of genetics and the bond of friendship: Bryon, who challenged me from day one and ensured that the worlds of literature and science would become a large part of my life. Your friendship and brotherhood are among the things dearest to me in my life. To Diana and the Dons: *Gracias por todo tu cariño y por proporcionarle tanta felicidad a la vida de mi hermano. Los códigos genético de nuestras dos familias ahora están combinadas para siempre con Tyler y Olivia.* And, of course, a special thanks to Tyler and Olivia, for keeping me real and reminding me, through repetitive example, of the supreme power of a child's smile.

To all my friends who have lent their talents to making this book the best it can possibly be: Paul Strachman, Nabil Kassam, Michael Novick, Bilal Musharraf, Matt Giudice, BJ Miller, Rico Cerrato, Mercedes Ruiz, Wonuk and Erina Lee, Marc Karimi, Amy Dosoretz, and Brian Fox. Thank you for your invaluable guidance in your respective fields of expertise and for being my true friends. And to Bethany Slater, one of the most talented surgeons in the world: Thank you for helping me stay focused for more than a decade and for being the never-ending source of pure beauty in my life. Always remember one thing: Cut here!

I was a professional student for a large part of my life, and I was fortunate enough to attend some truly amazing institutions and meet some exceptional people along the way. The person who has provided invaluable guidance from the very beginning is Dr. Gerald Acker: It is because of you that I ultimately took the path that I did, and I am forever grateful to you for being my mentor and my friend throughout.

A special thank-you to the Berlin and Guarino families, for your years of friendship, direction, and certitude.

To Dr. Robert Desnick, thank you for all of your invaluable guidance throughout my career in genetics. Working in your laboratory was all it took for me to understand that medical genetics would be a large part of my life.

To Dr. Stephen Hosea, a warm Kentucky thank-you for your tireless devotion to medical education and for teaching me and my fellow residents that the practice of medicine is as much an art as it is a science.

And a cardinal thank-you to Derrick Bolton, Garth Saloner, Bill Guttentag, Irving Grousbeck, Andrew Rachleff, and Stefanos Zenios at Stanford University for persistently asking what matters most to me and why and providing me with the skills I needed to turn my answer into reality.

Thank you to the University of Michigan and its Honors Program, Mount Sinai School of Medicine, Santa Barbara Cottage Hospital, and Stanford University's Graduate School of Business, the four noble institutions that have been instrumental in passing on the knowledge of our ancestors to me and millions of others. It is because of you that all of us who have been privileged to walk your halls are now empowered to change lives and change the world.

And to everyone at Santa Barbara Cottage Hospital: You have shown me that the practice of medicine is capable of achieving true excellence for physicians, patients, and the community all at the same time. My love of healthcare is very much a result of your brilliance and rationality.

To the best literary agent, Rick Broadhead: You believed in this project from the beginning and I sincerely appreciate all your hard work and effort. You went above and beyond and, throughout this process, have become a friend. Thank you.

To my editor, Judith Kern, who provided unyielding support and expertise throughout the entire process. Thank you for all of your insights, your ideas, your inquisitiveness, your creativity, and, of course, your exceptional editing skill. I truly do appreciate the time you spent making this book what it is now.

And thank you to Marian Lizzi, John Duff, Perigee, and the Penguin Group for your experience, fortitude, and understanding. I am grateful to you for providing such an excellent home for this book and for all the hard work you and your team have put forth on its behalf.

And, last but not least, thank you to all the genetics researchers worldwide who continue to work tirelessly to discover the exact genetic cause of disease. Your remarkable efforts have provided us with the ability we now have to outsmart our genes. Thank you—and keep up the amazing work!

I have heard there are troubles of more than one kind.
Some come from ahead and some come from behind.
But I've bought a big bat. I'm all ready you see.
Now my troubles are going to have troubles with me!

—Dr. Seuss

Contents

Introduction
An Invitation to Change Your Life

Imagine a world without disease. This world may one day be possible. How? Through scientific advances in genetic technology that allow physicians to look at your DNA and predict future disease. By predicting the diseases you are likely to encounter throughout your life as well as the diseases you may pass on to your children, genetic technology is now able to empower you and your doctor to fight against disease *before* it even manifests so that you can outsmart your genes and start protecting your future today.

The most powerful asset our species has to protect and preserve life is not our physical strength or speed or agility; it is our supreme ability to think. Eagles use their power of flight, speed, and eyesight to survive; lions use their size, muscular strength, and agility; and we humans use our minds, ingenuity, and fortitude. Although we may not be able to outmuscle or outrace some opponents, we have the ability to *outsmart* most adversaries, including disease. By allowing us to continually create, perfect, and benefit from technological advancements, our minds hold the key to our species' survival and well-being.

As the human lifespan continues to lengthen, we'll be faced with an increasing number of ailments and diseases. In the last 100 years, the average lifespan has increased from 35 to 80 years. Because of this, our society is now faced with an unprecedented number of diseases associated with aging, such as cancer, heart disease, and

Alzheimer's. Until now, our primary means of dealing with illness has been to wait until it appears and then try to treat it, often by drastic, debilitating, and painful means. But the best way to defeat disease will always be to avoid it altogether. We can do just that by studying our genetic code and using the information it provides to take control of our destiny.

Many of us go through life assuming that we're either blessed or doomed by our genetic inheritance. I'm sure you've heard people say things like, "I got my great memory from my mom," or "I can't lose weight; it's just in my genes. Both my parents are overweight." But the groundbreaking news is that, even if your genetic inheritance has put you at increased risk for a disease, science now has the means to help you *alter, minimize, and perhaps entirely avoid your current genetic destiny.*

With the completion of the Human Genome Project in 2003 and recent advancements in DNA testing and analysis, it is now possible to predict which diseases you are likely to contract so that you can take steps to lower and possibly eliminate your risk. The key is that almost all chronic disease results from a *combination* of genetic and non-genetic factors (such as what types of food you eat and your other lifestyle choices). Because of this, understanding your genes gives you the insight that is necessary to identify and selectively modify your nongenetic risk factors, thereby decreasing your overall risk of disease.

Traditional Western medicine has been criticized for being reactive rather than proactive because it usually doesn't take the offensive in attempting to prevent disease. Instead, Western medicine is primarily defensive because it tries to cure people of diseases they've *already* contracted. But with the new, revolutionary medical specialty called predictive medicine that no longer is true. The mandate of predictive medicine is twofold: to determine your personal genetic profile and then, most importantly, to provide you with the means of fighting off potential illness before it occurs. My purpose in writing *Outsmart Your Genes* is to bring predictive medicine to the attention of the general public—the countless individuals whose lives it could potentially save—so that they can partner with their physicians to harness its extraordinary life-changing power.

It was an accident of birth—that is, my own genetic inheritance—that first sparked my interest in genetics. I was born with a rare dominant genetic disorder called epidermolysis bullosa (EB for short), a condition that causes blisters to form when one's skin temperature rises above a certain level. Anything—friction on the feet caused by running or friction on the hands caused by gripping the handle of a tennis racquet, for example—that raises the skin temperature causes painful blisters to form that then need to be lanced with a scalpel, which can lead to potentially serious infections. Although my form of the condition is not as severe as others, it was, nevertheless, both physically and emotionally distressing for me as a child.

I remember asking my parents why I was different from all the other kids, and their answer was that I was different because of my genes. When I was older and had learned what the word *genes* really meant, I naturally asked my parents if anyone else in our family had EB. The answer was no. They had searched back through many generations and couldn't find a single family member with my condition. As I later learned, what this meant was that one of my genes had spontaneously mutated.

Because of my condition, I've had a deeply personal interest in genetics for my entire life. During high school biology I was introduced to the science of genetics, which was then still a relatively new and emerging field. Even then, however, I understood that genetics had the potential to create a paradigm-shattering advancement in science and medicine. If we could map and perhaps even manipulate our genes, it would be possible to conquer all kinds of diseases. I was hooked. Genetics would become my life's work.

When I entered the University of Michigan in 1996 there was no way to specialize in the study of genetics, but the university's Honors Program allowed me to create my own major, and, as a result, I am lucky enough to have been in the vanguard of genetic investigation ever since. While I was still an undergraduate, I conducted research in two different genetics laboratories—one at the University of Michigan and the other in the human genetics laboratory at the Mount Sinai Medical Center in New York City, which was, and still is, run by my mentor, Dr. Robert Desnick, chairperson, physician in chief, and dean of Mount Sinai's Department of Genetics and Genomic Sciences. Working in these labs, I began to see that the practical applications of genetics were virtually boundless.

After graduating from Michigan I entered the Mount Sinai School of Medicine, where I became even more aware of the ways genetics could affect medicine on the clinical level. Based on years of genetic research, the doctors at Mount Sinai had developed an enzyme-replacement therapy for the treatment of Fabry's disease, a rare and potentially fatal illness whose sufferers lack the enzyme that is responsible for breaking down an important type of fat found throughout the body. As a result, the fat accumulates and causes serious harm to the body's organs. The enzyme-replacement therapy, however, allows patients to live normal lives. Genetic research had successfully moved from the research laboratory to the patient's bedside and led to the control of a debilitating disease. It was clear that this was only the tip of the iceberg for genetics.

As a medical student and then as an intern, I saw many patients die of illnesses that would have been curable if they had been diagnosed sooner. People in their 20s, for example, were dying of heart attacks caused by a genetic predisposition they didn't even know they had because it produced no symptoms. The more I witnessed, the more I realized that the real key to creating a medical revolution was to change the existing medical paradigm from being reactive to being proactive, and the way to do that was by moving decades of genetic research out of the laboratory and into the hands of the people who needed it most: the patients and their doctors.

For the past few years, I've been working on ways to accomplish this by facilitating the integration of genetics into the practice of medicine. As a fully licensed medical practitioner and founder of both Existence Health, a predictive medicine practice, and Existence Genetics, a company that provides predictive medicine services to the healthcare industry, I conduct considerable research of my own and have become familiar with the tens of thousands of scientific studies on genetic testing. In this book I share with you the invaluable information provided by many of these studies so that you can better understand not only how genetic screening can benefit you right now but also how and why it is already changing the practice of medicine.

Technological advances are now providing ways for physicians to examine all of a person's genes at one time and at relatively little cost. What this means is that the gap that has long existed between genetics laboratories and physicians' offices can finally be bridged. Now we are able to link genetic predictions with actual medical practice. Predictive medicine has been born.

In the pages that follow, I first examine the extraordinary possibilities the power of predictive medicine gives you for outsmarting your genes. You'll see how science has taken us from the initial experiments of a 19th-century Austrian monk to our present-day understanding of how genes govern all the functions of our body. I'll also discuss our newfound ability to map the entire, unique genetic makeup of each and every individual, how this relates to you and your family, and the specific information you need to know before you undergo genetic testing. I will then explain in Part II how genes determine your risk for a wide variety of specific diseases and traits and, most important, *what you and your doctor can do right now*, based on the results of your genetic tests, to alter your genetic destiny.

Battling disease and suffering is my life's mission, and predictive medicine is the most powerful tool I can give you to maximize your health and prolong your life, no matter what your present age or health status. Acquiring and acting on the information I provide may well be a life-altering experience for you. The future is here, and I invite you to join me as we explore the many ways that predictive medicine can not only improve your health but potentially extend your life and the lives of those you love.

PART I

Let's Live Healthier— and Longer

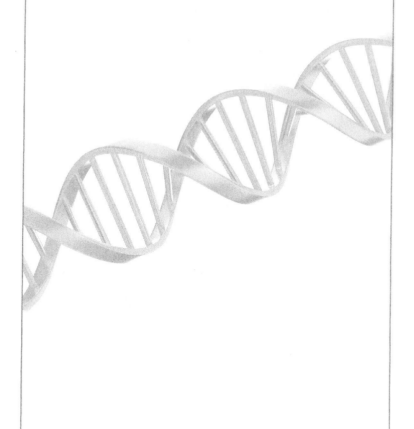

1

The Power of Predictive Medicine

MISCONCEPTION: There's no point to learning your genetic risk for a particular disease because there's nothing you can do about it.

FACT: If you discover that you are at increased risk for a disease, there are actions you can take to decrease the likelihood that you will get the disease or to limit its impact on you if it should ever manifest. And, in all cases, you will be alerted to diseases you might pass on to your children. If you know you're at risk, *there are things you can do about it!*

KNOW YOUR GENETIC MAKEUP AND TAKE CHARGE OF YOUR FUTURE

Your genetic makeup—that is, the genes you inherited from your parents—holds the secrets to your destiny. Not only the color of your eyes and hair and your musical or athletic abilities but also thousands of other factors related to your appearance, your health, and how you interact with the world are all determined, in whole or in part, by your genetics. Your genes determine how fast your metabolism works and how you process the calories you consume, which, in turn, determines how much fat you have around your abdomen and, to a significant degree, how much you weigh. Genetics are responsible for how your body processes medications, whether you'll experience side effects from a particular drug, what dose you require, and whether the drug will be effective at all. Your genetic makeup even determines your basic personality traits, such as if you're a risk taker, whether you are shy or outgoing, how you handle stress, and even if you're more inclined to keep anger bottled up or to let it out.

Some traits, such as eye and hair color, are determined entirely by genetic inheritance. For other characteristics, such as intelligence and height, your genes define the range of possibilities, and nongenetic factors, such as your education and lifestyle, determine where you ultimately fall within that range. For example, your genes may dictate that your adult height will be between 5 feet 8 inches and 5 feet 11 inches, but your actual adult height is also determined by nongenetic factors, such as the kind of nutrition you receive while growing up.

THE "GIVE ME YOUR LUNCH MONEY OR ELSE" GENES

MOST PEOPLE are surprised to learn that so much of what and who we are is predefined by our genes. Even characteristics that we think must be based purely on our environment are actually determined in part by our genetic makeup. As an example, you might be surprised to learn that a large scientific study published in 2008 in the *Journal of Child Psychology and Psychiatry* found that genes are about 60 percent responsible for why a child becomes a bully and about 75 percent responsible for why a child becomes a victim of bullying. The remainder is determined by nongenetic factors, such as the child's home environment.

Genes control many of our traits and characteristics, some of which—such as antisocial behavior, impulsive tendencies, and thrill seeking—make some children more susceptible to bullying, while other characteristics—such as being more introverted, more emotional, or having a tendency to cry easily—make other children more susceptible to victimization. Therefore, when a child who is predisposed to being impulsive, antisocial, and thrill seeking is around a child who is predisposed to being quiet and emotional, these children's genes have virtually set the stage for a bully–victim interaction.

Most important, your genes, either in whole or in part, determine whether you are at risk for specific diseases. Even your risk of contracting infectious diseases, ranging from the flu to HIV, is determined in part by your genes.

With the completion of the Human Genome Project and the advent of the powerful technologies now available for genetic testing, scientists have the ability to decode and analyze your genetic makeup and predict what diseases you are at risk for developing. But unless your genetic information is made actionable, your decoded genome is no more useful than a high-tech paperweight. Predictive medicine is the component that makes genetic testing actionable.

Predictive medicine is the component that makes genetic testing actionable.

Predictive medicine is a new medical specialty. When a physician believes that a patient needs a radiological examination such as a MRI or a CAT scan, he or she refers the patient to a doctor specifically trained in the field of radiology to perform the test. This test is then read by the radiologist, who supplies the referring physician with a written report. The report is what makes the test actionable for the physician—which is exactly what predictive medicine does in the field of genetics. *Any* physician can integrate predictive medicine services into his or her practice just as he or she does radiologic exams and laboratory tests.

The genetic report provides an analysis that clearly identifies the patient's risk for various diseases and specifies preventive measures that have been shown to decrease the risk or minimize the impact of those diseases. Even the preventive measures themselves can now be genetically tailored to your DNA. When your doctor receives your genetic report, he or she can work with you to take actions that will minimize your risk and perhaps even prevent you from ever developing the diseases for which you are at risk. Keep this in mind: Just because you have a gene that increases your risk for a certain disease does not necessarily mean you have or will ever contract the disease. What it does mean is that your genes *predispose* you to that disease.

But you might also find out that you *do not* have any of the genes associated with a particular disease—for example, if you find you do not have any harmful changes in the genes BRCA1, BRCA2, CHEK2, ATM, and FGFR2, all of which are linked to breast cancer, you might feel a significant relief. You may even discover that you have a beneficial genetic makeup that protects you from and *lowers* your risk of certain diseases.

One indication of the usefulness of comprehensive genetic testing in clinical care is evident from a 2009 study conducted by the director of the Centers for Disease Control and Prevention's Office of Public Health Genomics in Atlanta (also known as the CDC). This study, published in the journal *Genetics in Medicine*, found that when comprehensive genetic testing results for a patient were made available to the patient's physician, 75 percent of physicians changed some aspect of their patient's care "such as screening tests offered, medications or dosages prescribed, lifestyle changes recommended, frequency of follow-up appointments, or diagnoses made." Although the physicians were already treating these patients and had presumably taken a family history and conducted all other routine care, obtaining genetic information still made a difference in their clinical management of these patients *three out of four times*.

In terms of its impact on society, predictive medicine has the potential to significantly decrease the costs of healthcare. As just one example, consider the medication warfarin (Coumadin), which is one of the most widely prescribed drugs in the world and is used to thin a person's blood in order to protect against blood clots, heart attacks, and strokes. A senior member of the U.S.

> **Predictive medicine has the potential to significantly decrease the costs of healthcare.**

Food and Drug Administration's economic staff stated that if it became standard practice to conduct genetic testing before prescribing warfarin, the decrease in the number of adverse reactions would result in a net healthcare savings of as much as $1 billion per year.

The power of predictive medicine lies in its ability to look at your entire genetic profile and provide you with a forewarning that you and your physician can use to change your future health for the better. You, as a healthcare consumer, are the direct beneficiary of this revolution in medicine because what you can learn about your genes today will have a profound impact on your present and future well-being.

USE NATURE TO DETERMINE YOUR NURTURE

While genetics (often referred to as our nature) can be thought of as everything that affects you from the inside out, your environment (our nurture) is everything that affects you from the outside in and includes all nongenetic factors, such as how much you exercise, your level of education, your diet, and your medications. Because most diseases are determined by a combination of genetic and nongenetic factors, understanding your nature will empower you to make the changes in your nurture—such as lifestyle changes—that will minimize and potentially eliminate your chances of contracting the diseases for which your genes put you at risk.

Knowing your unique genetic makeup will also allow your doctor to provide guidance that is custom-tailored for you. He might recommend that you increase your consumption of specific foods, such as broccoli or fish, or decrease your consumption of others, such as salt or saturated fats. Your doctor might change or adjust your current medications, recommend particular tests (such as more frequent colonoscopies if you are at increased risk for colon cancer), or suggest alternative therapies such as yoga for stress reduction. Preliminary evidence has even shown that genetic testing can actually pinpoint which type of therapy will be best for treating someone with an alcohol addiction. With information like this, an addiction-recovery program can be tailored to the specific genetics of the sufferer.

Accepting the fact that genetics determines much of who we are is the first step toward outsmarting our genes and conquering disease. By factoring in our genetics we are not giving in to nature, rather we're being empowered to fight it.

Accepting the fact that genetics determines much of who we are is the first step toward outsmarting our genes and conquering disease.

THE PREMATURE LOSS OF A LEGEND

WHY SHOULD genetic testing matter to someone who is healthy, whose parents are healthy, whose children are healthy? The answer is perhaps best conveyed by the story of Olympic figure skating champion Sergei Grinkov.

Sergei and his wife, Ekaterina Gordeeva, were among the most celebrated pair skaters of all time. They met in Russia when Sergei was 14 and Ekaterina (Katia) was 10, and the pair won four world championships, two Olympic gold medals, and the hearts of millions of fans. Sergei did not smoke, he didn't use drugs, he ate in moderation, he exercised for several hours every single day, he had no history of diabetes, and all his medical exams indicated that he was an extremely healthy athlete. He did, however, have high blood pressure, and his father died suddenly from a heart attack at the age of 52, but because Sergei was so young and athletic, these two warning signs didn't sound any alarms for his doctors.

On November 20, 1995, when Sergei was 28 years old, he and Katia went to the ice rink as usual to practice. As Katia wrote in her 1996 autobiography, *My Sergei: A Love Story*:

Sergei was gliding on the ice, but he didn't do the crossovers. His hands didn't go around my waist for the lift. I thought it was his back. He was bent over slightly, and I asked him, "Is it your back?" He shook his head a little. He couldn't control himself. He tried to stop, but he kept gliding into the boards. He tried to hold onto the boards. He was dizzy, but Sergei didn't tell me what was happening. Then he bent his knees and lay down on the ice very carefully. I kept asking what was happening. "What's wrong, Serioque? What's the matter?" But he didn't tell me. He didn't speak at all.

Sergei died that day, leaving behind a wife, a 3-year-old daughter, and an unfinished life. Researchers from the Johns Hopkins School of Medicine were so interested in why a seemingly healthy 28-year-old athlete would die so suddenly that they obtained permission from Katia to conduct posthumous genetic testing on her husband. One of the researchers, Dr. Pascal Goldschmidt-Clermont, found that Sergei had a genetic variant that a previous study had linked to an increased risk of early-onset heart attack. If Sergei had known that his genetic makeup contained this abnormality, he and his doctors could have instituted a wide range of interventions, and his condition would have been closely monitored by a cardiologist. It's true that he did have at least a couple of warning signs: high blood pressure and a father who died young of a heart attack. But as is often the case with a seemingly healthy person of Sergei's age, the warning signs were overlooked.

THE GENETIC ANSWER TO GENERIC RISK FACTORS AND FAMILY HISTORY

People who fall into particular categories based on gender, age, ethnicity, and lifestyle habits are assumed to be more or less predisposed to contracting particular illnesses. These are *generic risk factors* that your doctor takes into account when diagnosing a complaint or determining what to look out for. If you are a woman past menopause, for example, your doctor knows you are at greater risk for heart disease than a woman who is still menstruating. If you smoke, you are at greater risk for lung cancer than someone who never smoked. But the problem with depending on these nonpersonalized generic risk factors is that *they* are generic, but *you* are unique. You are fundamentally different on a genetic level from all other people, even if they share your gender, age, ethnicity, and lifestyle.

Looking at your family history may offer a somewhat more accurate picture of your predisposition for particular diseases, because members of your immediate family do share *some* of your genes, and they often share similar habits and lifestyle. But family history is an imperfect science that, at its best, is able to provide only limited actionable information.

The quality of the family history obtained is also very physician dependent. Many physicians just don't have the time to go through a very thorough family history with all of their patients. Instead, patients are asked to fill out a questionnaire that asks about any family history of heart disease, diabetes, cancer, and a few other illnesses. If the patient isn't sure or marks one of these as no, many times the physician will not make any further inquiries. While medical students are taught the importance of taking a thorough family history to determine significant areas of inquiry and investigation, such thoroughness is seldom practiced in the real world of clinical medicine.

With the availability of comprehensive genetic testing, your doctor no longer has to rely on your memory, anecdotal evidence, or a diagnosis made decades ago that may or may not have been accurate to predict your future health. Generic risk factors and family history provide, at best, hints of the diseases for which you might be at risk. Genetic testing provides clear, objective, and personalized scientific answers. And,

With comprehensive genetic testing, your doctor no longer has to rely on your memory, anecdotal evidence, or a diagnosis of a relative made decades ago to predict your future health.

if you do not have very much information about your biological family, genetic testing may be your *only* means of finding out what diseases you're at risk for.

While generic risk factors and family history are useful as *adjuncts* to comprehensive genetic testing, they are not a substitute for it. The three used in combination provide much more information than any one used alone.

WHAT GENETIC TESTING CAN DETECT

Genetically speaking, there are three categories of disease. The first comprises common diseases, which are determined by a *combination* of genetics and nongenetic factors. If you have harmful changes in a gene for one of these diseases, you will be at increased *risk* for getting the condition in the future, but that doesn't mean you definitely will.

The last two categories are those that depend solely on your genes, with no nongenetic component. These are rare diseases and may be either dominant or recessive. A disease can be dominant or recessive because we all have two copies of every gene. If you have at least one copy of a gene for a dominant disease or if you have two for a recessive disease, you will most likely have the disease. If you have only one copy of a gene for a recessive disease, you will not have the condition; you are what is known as a carrier and you'll have the potential to pass on that disease-associated gene to your children.

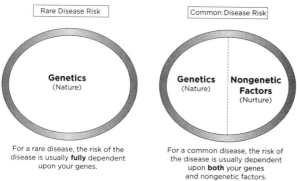

| Rare Disease Risk | Common Disease Risk |

Genetics (Nature)

Genetics (Nature) **Nongenetic Factors** (Nurture)

For a rare disease, the risk of the disease is usually **fully** dependent upon your genes.

For a common disease, the risk of the disease is usually dependent upon **both** your genes and nongenetic factors.

Through the use of comprehensive genetic testing, predictive medicine can provide you with highly actionable information that will *decrease* your risk of future disease, *reduce* the effect and severity of diseases you already have, and *protect* your future children against diseases you may carry.

CATEGORY 1: RISK FOR FUTURE DISEASE

Genetic testing can discern whether you are at an increased or decreased risk for certain diseases *in the future*. These are primarily common conditions, and their risk factors depend on a combination of genetic and nongenetic factors. Hundreds of diseases, including obesity, cancer, heart disease, Alzheimer's, Parkinson's, multiple sclerosis, Crohn's, macular degeneration (the leading cause of adult-onset vision loss and blindness), arthritis, and addiction. Genetic testing can even determine your degree of susceptibility to infectious diseases such as HIV/AIDS, malaria, and stomach flu. Other characteristics that have a genetic component are your response to medications and your intelligence, height, and athletic performance. Even male-pattern baldness is the result of a combination of genetic and nongenetic factors.

The most important piece of information to remember is that even if your genes predispose you to a disease, your total risk of contracting the disease is *not* set in stone. With information gathered from genetic testing, predictive medicine can help you take measures that will reduce your overall risk, outsmart your genes, and perhaps keep you from ever getting the disease.

Even if your genes predispose you to a disease, your total risk of contracting the disease is *not* set in stone.

In summary, your risk for common diseases can be shown in the following equation:

Genetic risk + Nongenetic risk = Total risk of disease

If you decrease your nongenetic risk, even though your genetic risk remains unchanged, your total risk of contracting the disease will decrease.

The best chance you have of beating disease is never to be faced with it, and the best way for you to avoid disease is through gaining the foreknowledge provided by genetic testing and predictive medicine. By predicting your future, you are empowered to change it.

By predicting your future, you are empowered to change it.

CATEGORY 2: DISEASE THAT AFFECTS YOU RIGHT NOW

Some diseases are 100 percent genetic, and if you have one dominant or two recessive genes for one of these diseases you *will* most likely be affected to some degree. By becoming aware of these diseases as early as possible, you can implement measures that will allow you to live a longer, healthier life.

You might think that if you had a disease you'd certainly know about it, but the fact is that many diseases have no symptoms whatsoever for years or even decades. Or you could have a disease whose symptoms are nonspecific and, therefore, might not obviously point to an exact cause. For example, diseases that can affect the electrical conduction system in your heart (which is absolutely essential for life) may show no symptoms at all until they cause sudden death.

A disease called malignant hyperthermia is completely dormant until you are exposed to anesthesia, at which point you may have a severe reaction that leads to death on the operating room table. Or you may have a genetic problem with your muscles that causes you to feel exhausted after even a small amount of exercise. Many people with this problem are overweight and are, therefore, told to exercise more, when in fact they are genetically incapable of exercising.

If genetic testing shows that you have one of these insidious diseases, you and your doctor will know what you are up against and will be able to treat it appropriately. For example, heart arrhythmias can be monitored in many different ways with various medical devices and controlled with medical procedures and medication. Malignant hyperthermia can be prevented by avoiding specific types of anesthesia commonly used during surgery. And if you are overweight because your genetic makeup causes you to be intolerant to exercise, you can focus on other weight-loss interventions such as diet, appetite-suppressant medications, and even surgical procedures such as gastric-banding or gastric bypass.

The essential first step to treating a disease, particularly when it is asymptomatic, is to know that you have it—through genetic testing.

CATEGORY 3: DISEASES YOU CARRY THAT DO NOT AFFECT YOU

Genetic testing for recessive diseases, those that usually require two copies of a disease-causing gene, will reveal your carrier status. If you have a single copy of one of these recessive genes but do not have the disease, you are considered to be a "carrier" of the disease. There are thousands of these rare recessive diseases, including Tay-Sachs, sickle-cell anemia, and cystic fibrosis, that fall within this category. If you carry a single gene for one of these diseases you will *not* be affected by the disease, but you can still pass the gene along to your children. And if your child's other parent is also a carrier of the same recessive gene and also passes it along, the child may get the disease.

The figure at right depicts the probabilities that a child born from parents who are both carriers of a recessive disease will inherit the disease. Each and every child will inherit, at random, either the disease-carrying or the unaffected gene from the father and either the disease-carrying or unaffected gene from the mother. Each of the four squares represents one of the four possible combinations. As you can see, if both parents carry a recessive gene for a particular disease, there is a 25 percent chance that their child will have the disease (lower right square), a 50 percent chance that he or she will be a carrier, and a 25 percent chance that he or she will neither have the disease nor be a carrier (upper left square).

		Father's Genes	
		Unaffected Gene	**Diseased Gene**
Mother's Genes	Unaffected Gene	Unaffected Gene Unaffected Gene (Noncarrier)	Unaffected Gene **Diseased Gene** **(Carrier)**
	Diseased Gene	**Diseased Gene** Unaffected Gene **(Carrier)**	**Diseased Gene** **Diseased Gene** **(Diseased)**

Recessive Inheritance

If you have the foreknowledge that you and your spouse both carry a disease, you'll be able to use that information to consider the multitude of family planning options that will significantly reduce the risk of your future children being afflicted by the disease. However, because you don't actually have the disease, the only way to determine if you are a carrier and protect future generations is to have genetic testing before you start a family. We'll discuss genetic testing for prospective parents in greater detail in Chapter 6.

IS GENETIC TESTING RIGHT FOR YOU?

Some people are fearful of genetic testing and say, "I'd rather not know." Others are more accepting of this new technology and, without much discussion, say, "Test me for everything!" The approach I advocate is somewhere in the middle. By avoiding all genetic testing you could be missing out on potentially life-saving information, but testing for all diseases at once has the potential to create information overload.

In terms of the fear factor, the information provided by genetic testing is no different from the information you or your physician derive from any other form of medical testing. Compare it, for example, to having your blood pressure taken. If you have high blood pressure you're at increased risk for many different diseases, including heart attack and stroke, and your doctor will take preventive measures, such as counseling you on lifestyle modifications, prescribing medication, and making sure to closely monitor your pressure thereafter. As a society, we are quite comfortable with screening for high blood pressure and using that information to alter behavior, and I believe that in time we will become just as comfortable with genetic testing.

Genetic testing is actually quite similar to wearing a bicycle helmet. No one ever plans to get into an accident while riding a bike, but if they do, the helmet can mean the difference between life and death. Just as a bike helmet protects you against a *potentially* deadly impact, genetic testing gives you the ability to avoid or considerably limit the harmful effects of a *potentially* life-threatening disease.

Of course, discovering that your genes predispose you to a disease doesn't help unless you do something about the results. In medical school we are taught never to run a test on a patient unless we plan to act on the results. If the test is not going to have any effect on patient care and if it is not going to be actionable, the test should *not* be ordered. The same philosophy can be applied to predictive medicine: Don't pursue genetic testing unless you and your physician are going to take action based on the results, such as by implementing preventive lifestyle changes.

MEDICINE'S NEW PINCER MOVEMENT

GOING ON THE OFFENSIVE AGAINST DISEASE

One of the main criticisms of traditional Western medicine is that its main goal is to cure a disease once it's taken hold rather than to prevent the onset of disease in the first place. Predictive medicine is now changing that approach and creating a revolutionary medical paradigm. Fighting disease is like fighting an ongoing war, and predictive medicine provides a new and effective battle strategy.

Fighting disease is like fighting an ongoing war, and predictive medicine provides a new and effective battle strategy.

One of the most highly regarded military strategies of all time is known as the pincer movement. The strategy involves attacking the enemy's flanks simultaneously and, in essence, encircling and entrapping the enemy. In terms of predictive medicine, we can now simultaneously attack both the genetic and nongenetic factors that can lead to disease.

Traditional Strategy

Disease Enabling Genetic Factors　　**Disease Enabling Nongenetic Factors**

Disease

Head-On Battle

Traditional Medicine

Healthcare

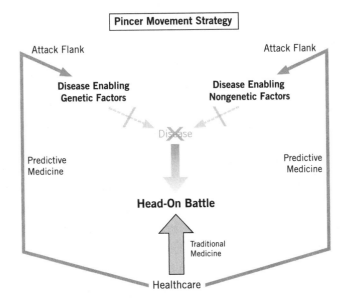

Without the information provided by genetic testing, traditional medicine has been primarily confined to fighting disease head-on after it manifests, which means that our medical artillery has always been defensive. Now, empowered by predictive medicine, we can finally take the offensive. We can use genetics both to prevent disease from happening and to tailor the most effective medical treatments to prevent the disease from winning should it occur. When traditional and predictive medicine join forces, they significantly increase your chances of conquering the enemy—disease.

2

From Peas to Predictive Medicine

MISCONCEPTION: Since the completion of the Human Genome Project in 2003, there's nothing more to learn about our genes and how they control our destiny.

FACT: Since 2003, tens of thousands of scientific studies have provided actionable insight into how our genetic makeup either increases or decreases our risk for disease.

Forrest Gump famously told us that his momma always said, "Life is like a box of chocolates; you never know what you're gonna get." As it turns out, an Austrian monk named Gregor Mendel had the very same thought back in the 1860s when he was working in a garden at a monastery in what is now the Czech Republic. Unlike Forrest Gump's mother, however, Mendel concluded that it *was* possible to know what you were going to get. In fact, it was his groundbreaking experiments that proved we could predict how life would turn out.

Mendel was fascinated with something most of us consider quite ordinary: the pea. So while the other monks at his monastery ate peas, Mendel began to study them. He observed that some peas had smooth skin while others were wrinkled, and some were yellow while others were green. He began to wonder why some peas looked different from others, whether there was any discernible pattern to those characteristics, and whether he could *predict* the odds of specific traits showing up in the offspring of two adult pea plants. He wanted to determine whether we could actually know what we were going to get before we got anything, and to do that he began to experiment with peas in the monastery's garden. He crossed smooth-pea-producing plants with wrinkled-pea-producing ones, green-pea-producing plants with yellow-pea-producing ones, and so on, and he recorded what the progeny of each cross looked like, generation after generation. In the end, he studied some 30,000 pea plants.

Mendel was the first person to describe the invisible factors that control the way traits are passed down from parents to offspring. What he was really describing for the first time were genes. Some of these factors, he found, could be considered *dominant* while others were *recessive*. When the dominant factor, such as green color, is present, it covers up, or masks, the recessive yellow color. Yellow peas, therefore, could not contain any green factors at all; if they had even a single green factor, they'd be green. But if they lacked all green factors, they'd be yellow. Through his experiments, he was not only deciphering inheritance patterns but also determining ways to predict the odds of obtaining certain traits in future generations. Mendel's discoveries clearly showed that if life is like a garden of pea plants, then, yes, we really can predict what we're likely to get. These early discoveries formed the basis for what has now become a revolutionary new approach to healthcare—predictive medicine.

PEAS VS. BEES

AFTER CONSUMING so many peas, the other monks at the monastery were probably not too upset when, in 1863, Mendel moved on from peas and started studying honeybees. The food at the monastery must have become a lot more enjoyable after that, but Mendel's new obsession soon turned bittersweet.

In the course of his experiments, Mendel mated honeybees from Egypt with honeybees from South America. The offspring of these pairings produced extraordinarily sweet honey, but a painful problem soon became overwhelmingly apparent. The hybrid bees turned out to be extremely ferocious, and they started to sting everyone—not only the monks at the monastery but also villagers for miles around. Because of this, Mendel decided to stop his experiments and destroy the hives. In the end, the sweetness of those bees turned sour, and plain old peas must have started to seem quite tasty.

In 1866, Mendel published a paper on his theories of inheritance and sent copies to prominent scientists throughout Europe. An unproven theory asserts that he even sent a copy to Charles Darwin, who supposedly never read it. Mendel turned out to be ahead of his time, and European scientists ignored his findings for decades. It wasn't until long after his death, at the dawn of the 20th century, that his long-forgotten papers were rediscovered and his groundbreaking work was finally recognized. Because of Mendel, scientists knew that offspring of virtually every species, from the simplest to the most complex, inherit traits from their parents in statistically predictable patterns. Gregor Mendel is now considered the father of modern genetics, and the basic principles of genetics are still known as Mendel's Laws of Inheritance.

Over the next 100 years, scientists from all over the world worked to solve the mystery of how this genetic inheritance occurs. They discovered that:

- Deoxyribonucleic acid (DNA) is responsible for inheritance. DNA is composed of only four chemicals that are commonly referred to by the first letter of their chemical name. These four chemical "letters" exist in various combinations, and the exact sequential order of those letters is known as the genetic code. Humans have 6 billion letters in their entire genetic makeup.

- Genes are the basic units of inheritance and are composed of DNA. Since genes are composed of DNA, each gene is made up of a series of chemical letters.
- A complete set of genes is known as a genome. Humans have approximately 20,000 genes in their genome.
- Genes exist in a linear order, one after another, on long thread-like structures of DNA called chromosomes.
- Almost all the cells in our body contain 23 *pairs* of chromosomes, meaning that each cell actually has a total of 46 chromosomes. And because genes exist on chromosomes, each of those cells also has two copies of each gene.
- Changes in our genetic makeup can occur. A change in one or more letters of the genetic makeup is called a "genetic variant," also commonly referred to as a "mutation."
- The specific combination of variations in the genetic code is responsible for our traits and predisposing us to diseases. Different people have different combinations of variations in their genetic makeup, which is why people have different traits and get different diseases. A change in just a single letter out of the 6 billion letters in our genome can cause disease. Genetic testing is performed in order to detect such variation.

If we compare genetic terminology to a book, we can think of genetics as being the *Instruction Manual of Life.*

GENETIC TERMINOLOGY	ANALOGOUS TO OUR *INSTRUCTION MANUAL OF LIFE*
DNA	The ink
Genetic makeup	The letters
Genes	The sentences
Chromosomes	The chapters
Genome	The entire book
Genetic variant	A typo
Genetic testing	Proofreading each letter

By the mid-20th century, scientists discovered that genes exert their effects on cells by coding for proteins. What this means is that the specific chemical letters of each gene can be translated by cellular machinery into a protein. This is analogous to Morse code, which encodes information using rhythms. For example, in Morse code the rhythm di-di-di-dah-dah-dah-di-di-di (also referred to as dot-dot-dot-dash-dash-dash-dot-dot-dot) codes for the distress signal known as SOS. The information is encoded in the rhythm, which can then be translated into understandable language by anyone who knows the code. The same is true of a gene: Its information is encoded in its chemical letters, and the cell is able to translate this code into a protein. That's why we say that each gene *codes for* a specific protein. Because proteins are responsible for building all of the cells of our body and for directing the majority of cellular functions, the fact that genes code for proteins means that genes contain the code of life.

Your genes control, either in part or in full, all of the processes of your body and variations in your genes are responsible for predisposing you to diseases.

In terms of predictive medicine, the most important concept is that your genes control, either in part or in full, all of the processes of your body and that variations in your genes are responsible for predisposing you to diseases. Therefore, by analyzing your genetic makeup we can predict the diseases you're likely to face in the future and tell you how to best avoid them.

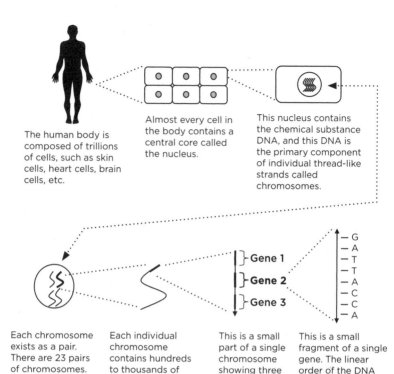

The human body is composed of trillions of cells, such as skin cells, heart cells, brain cells, etc.

Almost every cell in the body contains a central core called the nucleus.

This nucleus contains the chemical substance DNA, and this DNA is the primary component of individual thread-like strands called chromosomes.

Each chromosome exists as a pair. There are 23 pairs of chromosomes.

Each individual chromosome contains hundreds to thousands of genes.

This is a small part of a single chromosome showing three genes.

This is a small fragment of a single gene. The linear order of the DNA letters of the gene is the genetic code.

EXAMPLE OF THE GENETIC CODE

- G
- A
- T
- T
- A
- C
- C
- A

EXAMPLE OF A GENETIC VARIANT*

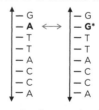

version 1   version 2

The sequence of DNA letters in your genes is translated by the cell into proteins. These proteins build and run the cell. Because of this, your genetic code is responsible for dictating the everyday functioning of all the cells in the entire body.

A change in just one of the 6 billion letters of your genetic code can lead to a change in the protein produced by a gene. This abnormal protein can then go on to affect the overall cell. Abnormal cells can then affect your entire body, potentially causing illness and disease.

Today you are You,
that is truer than true.
There is no one alive who is Youer than You.
—Dr. Seuss

ALL HUMANS have a total of 23 pairs of chromosomes in each cell. The only cells in your body where chromosomes do not come in pairs are sperm and egg cells, which contain only a single copy of each chromosome. Because conception occurs when a sperm and an egg come together and combine to form an embryo, we get half our chromosomes from our mother (via the egg) and half from our father (via the sperm), which is why each successive generation of offspring always has the same total number of chromosomes.

What makes each of us unique is the randomness by which the particular chromosome of each pair is contained within a particular egg or sperm cell. Because of this randomness, no two eggs and almost no two sperm will ever have exactly the same combination of chromosomes.

Here's how it works. All your cells (except the egg and sperm cells) have two chromosome 1s (call them 1A and 1B), two chromosome 2s (2A and 2B), two chromosomes 3s (3A and 3B), and so forth, all the way up to chromosome 23. Although both chromosomes in any given pair contain the same genes, the actual genetic code within those genes may be slightly different. Therefore, at the level of the genetic code, each chromosome in a pair is actually unique and contains different information.

Here's a simplified way to conceptualize just how unique each of us is on a genetic level: Because sperm and eggs contain only a single version of each chromosome and because the specific chromosome (A or B) of each pair that ends up in an individual sperm or egg is completely random, the number of

continued

possible combinations of chromosomes in a single sperm or egg cell (such as 1A, 2A, 3A . . . 23A or 1A, 2B, 3A . . . 23A or 1A, 2B, 3B . . . 23B) is 2^{23}, or more than 8 million possibilities!

Because a woman produces only around 400,000 eggs in her lifetime, it is likely that every single egg is unique. Because a man produces approximately 12 trillion sperm in his lifetime, more than one sperm will carry the same exact genetic makeup. However, because each person results from the combination of one egg and one sperm, the odds of two different sperm and two different eggs from the same parents creating two children with the same exact genetic makeup is less than 1 in 70 trillion ($2^{23} \times 2^{23}$).

That's why you're different from your mother, father, brothers, and sisters—while you each received 23 chromosomes from each parent, the exact combination of chromosomes and the exact combination of genes and variants on each chromosome are unique, and this exclusive combination makes you, you! Because it's estimated that over 100 billion humans have walked the earth, unless you are an identical twin, you can safely say there has never been anyone like you: You—just like the rest of us—are 1 in 70 trillion and, therefore, you are truly one of a kind!

DISCOVERING THE STRUCTURE OF DNA

Building on the work done by other scientists before them, James Watson and Francis Crick, who met at Britain's Cambridge University in 1953, were able to unravel the structure of the DNA molecule and solve one of the major mysteries of inheritance.

DNA, it turned out, is a double-stranded helix that coils round and round like a spiral staircase. The two strands are connected by rungs (just like the rungs of a ladder) that are composed of four chemicals: adenine, cytosine, guanine, and thymine, which are generally referred to simply by the first letter of their names (A, C, G, and T). These chemical letters represent the only four letters in the entire genetic alphabet. Amazing as it may seem, your entire genetic makeup, which is the instruction manual for your life, is composed of only these four chemical letters in various combinations.

THE DOUBLE HELIX STRUCTURE OF DNA

Besides describing the structure of the DNA molecule, Watson and Crick also suggested how the molecule works and how it replicates so that genetic information can be passed down from generation to generation. Just before a cell divides, the two sides of its DNA helix unzip (the rungs split down the middle, and the two strands of the helix disconnect from one another) so that each strand acts as a template for making an exact copy of the original genetic material. Therefore, each time DNA unzips and replicates, you get two copies of the original. When a single cell wants to divide into two daughter cells, one of the copies goes to one daughter cell and the other copy goes to the other so that no cell is ever left without a complete set of instructions.

Building on the foundation laid by Watson and Crick, scientists from around the world started to conduct considerable amounts of research in order to unravel the remaining secrets of DNA. Now that they knew what the forest looked like, they started examining the individual trees. One of the greatest discoveries in this regard came in 1961, when Marshall Nirenberg finally cracked the actual genetic code. While all the scientists who had gone before him had clarified the big picture, Nirenberg figured out how to read the code that made up each individual gene. Going back to our earlier analogy, Nirenberg's discovery was tantamount to figuring out how to decipher Morse code without the aid of a user's manual. In essence, he cracked the code of life.

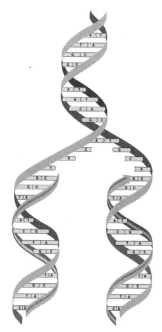

Replication of DNA from one copy into two.

GENETIC VARIANTS—THE GOOD, THE BAD, AND THE NEUTRAL

Although most genetic variants have no effect on the person who inherits them, some can be beneficial and others detrimental to health.

Variants within the genetic code of the BRCA1 gene, for example, are known to be associated with an increased risk for breast cancer, although the BRCA1 gene itself (if it doesn't contain any variants) actually protects against breast cancer. In other words, it's the variants within a gene that are to blame for causing disease, not the gene itself. In general, all our genes have evolved naturally over tens of thousands of years to promote and increase our ability to survive. As a gene is passed down from generation to generation, variants can arise spontaneously and these variants can alter the function of that gene—because of this, the genes can change over time and this forms a cornerstone of evolution. But sometimes a change *within* a gene can cause it to act in a way that is *detrimental* to our health, which is why particular genes are now known to be associated with particular diseases.

Many factors can cause genetic variants. Some are brought about by excessive sunlight, environmental pollution, smoking, radiation, poor nutrition, and other environmental and lifestyle choices. Others occur with aging or through occasional errors, such as when a cell divides.

To understand how a variation within DNA can occur, you might want to think of it in terms of your transcribing a long document by typing the text into your computer. You'll probably make a few typing errors. (In fact, there's probably never been a book published that didn't have at least one typo in it.) It's almost inevitable, which is one of the reasons why every person's genetic makeup (which consists of 6 *billion* letters) contains a number of variants. Because you have trillions of cells throughout your body and a small percentage of them divide every day, scientists believe that several thousand errors occur on a daily basis. Usually (but not always) when a variant occurs, your body detects and corrects the error, just as your computer often—but not always—detects and automatically corrects your typing errors. But even if your body doesn't make the correction, the variant is most often harmless and does not affect your health.

When a variation occurs in a sperm or egg cell that is fertilized and becomes an embryo, the variant becomes a permanent part of the resulting child's DNA and exists within each one of his or her cells, including half of his or her sperm or egg cells. (It's 50 percent because, as we've discussed, even though you have two full copies of your entire genetic makeup in every cell, the single copy that is contained within each sperm or egg cell is random, so if a variant occurs in one of the two copies, each sperm or egg cell has a 50 percent chance of containing it.) When that child, in turn, grows up and has children, they too will have a 50 percent chance of inheriting that variant. This is how genetic variations become a permanent part of our DNA and get passed on from generation to generation. In fact, it is how evolution occurs.

It's important to understand that, because everyone's DNA contains genetic variants, there is no such thing as a perfect genetic makeup. Perfection does not exist within genetics because it is imperfection that leads to diversity, and diversity is what enables us to evolve, adapt, and survive. Without imperfections (that is, variants) occurring in DNA, we wouldn't be here today. However, while some of the imperfections (genetic variants) have beneficial effects, others are harmful, and it is these harmful variants that we can now start to detect and outsmart with genetic testing and predictive medicine.

DECODING ANCIENT DNA

USING ADVANCED techniques, scientists can now estimate the date when a specific genetic variation first occurred. For example, a variant in the *MYH* gene has been found to have occurred about 5.3 million years ago, even before our species, *Homo sapiens*, existed, and this variant caused the facial muscles involved in the chewing of food to become smaller. This was important because, before that time, the facial muscles were so large that they took up a significant portion of our distant ancestor's (the chimpanzee) head and, therefore, limited the size of its brain. The smaller facial muscles resulting from this genetic variation meant the brain had room to evolve and grow significantly in size. This particular bit of evolutionary history is confirmed by paleontological specimens that show a gradual increase in brain size over a long period of time until part of the chimpanzee line eventually evolved into a separate and distinct species—the human!

You may remember that at the beginning of Michael Crichton's best-selling book and blockbuster movie *Jurassic Park*, scientists retrieved dinosaur DNA from a mosquito that had become trapped in tree resin millions of years ago. This is actually scientifically plausible because DNA is quite stable over long periods of time, especially when it is frozen in permafrost or encapsulated in a substance (like tree resin) that limits exposure to air. In fact, scientists have already studied the DNA of Egyptian mummies and even that of ancient species, including Neanderthals, who became extinct 30,000 years ago. What we've found is extremely interesting. From sequencing the

Neanderthals' genome we now know, for example, that they were lactose intolerant; that some of them had red hair; and that they shared some of our genes associated with speech, so they may have had a rudimentary language.

One day in the not-so-distant future we might actually be able to resurrect extinct species, such as dinosaurs and Neanderthals, by using ancient DNA. This possibility is now much closer to reality than to science fiction.

For every disease associated with genetic variants that increase one's risk of the disease, there are usually other variants that can decrease that risk or impart special traits. People who have these variants are the true "X-Men" of our species—and you may very well be one of them. Here are a few of these remarkable variants and their effects.

EXTRAORDINARY TRAITS		
TRAIT	**GENE(S)**	**HOW IT WORKS**
Exceptional longevity	*CETP, APOE, FOXO3A*	These genes control the way the body processes cholesterol. In certain populations, variants within these genes cause people to have either higher high-density lipoproteins (HDL, the good cholesterol) and lower low-density lipoproteins (LDL, bad cholesterol) levels or increased protection against oxidative stress at the cellular level. These result in a decreased risk of cardiovascular disease that greatly increases the person's chance of living to 100 years old and longer. Some of these genetic variants also appear to preserve the mental faculties, so people who have them not only live much longer but also stay mentally sharp.
Great physical strength	*MSTN*	This gene codes for myostatin, a protein that *limits* how much muscles can grow. A variant within this gene causes the body to produce less myostatin, leading to Schwarzenegger-like muscle mass and superhuman strength, even in babies.

continued

Enhanced memory	*WWC1*	This gene is very active in areas of the brain that are associated with memory. A variant in this gene allows the brain to retain memories with less effort. Because of this, people with this variant have much better short-term and long-term episodic memory. Preliminary evidence also suggests that this same variant may protect against Alzheimer's disease. Although the variant occurs in peoples from around the world, it is most frequent in Asian populations.
Elite athletic performance	*ACTN3, ACE, EPAS1*	These genes produce important proteins in our muscles. Variants within these genes affect the way the muscles use energy when they are working hard during exercise (known as muscle efficiency) as well as how much strength the muscles can gain through training. Other variants in these genes also affect the body's oxygen-sensing mechanisms, such as how the lungs and heart deliver oxygen to the muscles during exercise. Some of these variants are associated with a predisposition for short, power exercise, such as soccer, sprinting, short-distance swimming, gymnastics, tennis, boxing, wrestling, and weightlifting. Other variants are associated with a predisposition for long, endurance exercise, such as running marathons, cross-country skiing, long-distance swimming, long-distance biking, mountain biking, triathlons, rowing, and mountaineering. Studies have also looked at elite athletes, including many Olympians, and have found that many of them have specific patterns of variants within these genes. (Optimizing athletic training through genetic testing is discussed further in Chapter 5.)
Increased IQ with breastfeeding	*FADS2*	Breastfeeding has been shown to increase the lifelong IQ of the child, but only if the child has a specific genetic variant within the *FADS2* gene (see Chapter 6).
Intelligence and cognitive ability	*CHRM2*	Although the study of the genetic basis for intelligence is very controversial, it has been shown conclusively that cognitive ability and IQ are determined by both genetic and nongenetic factors, such as education level. The genetic component constitutes between 50 and 80 percent of IQ; the rest is determined by nongenetic factors. This gene produces a

Intelligence and cognitive ability	CHRM2	protein that is involved in how the brain processes and transmits information (called neurotransmission). Specific variants within this gene have been associated with a difference of about five IQ points.
Resistance to HIV infection	CCR5	This gene is responsible for producing a protein that resides on the outside of the immune cells (also known as white blood cells). When a person is exposed to HIV and the virus enters the blood, it infects the immune cells by clinging to the protein. Once the virus is able to infect a person's immune cells, it can replicate and spread, and the person then becomes "HIV positive." Therefore, this gene makes the protein that is necessary for HIV to infect a person. However, a specific genetic variant within this gene prevents it from working, which means that the protein isn't ever produced. If the HIV virus doesn't have this protein to cling to, it cannot infect the person and eventually the virus is cleared from the person's body, leaving the exposed person *uninfected* by HIV. As many as 1 out of every 100 people of European decent have two copies of this variant, with the frequency being lower in other ethnicities. (Further discussion of how genetic testing empowers us in our fight against HIV can be found at www.Out smartYourGenes.com/HIV.)
Protection against obesity	Many different genes	Many different genes are involved in the regulation of metabolism and the way the body processes and stores the calories that are consumed. Some contribute to weight, others to body mass index (BMI), and some dictate the amount of fat stored in different areas of the body. Variants within these genes have been found to be associated with leanness (protection against obesity), and others are associated with weight gain, leading to obesity. What's more, there are actually genetic variants that dictate whether eating a low-carb or low-fat diet will be most beneficial for losing weight. (Obesity is discussed further in Chapter 5.)

Protection against male-pattern baldness	AR, 20p11	Just by looking at pictures of the members of some families, you would probably guess that male-pattern baldness has a significant genetic component. Scientists have known for a long time that men with male-pattern baldness have an increased number of receptors for the hormone testosterone (a receptor is used by the cell to catch and bind to a substance, such as testosterone, as it circulates through the body), but only in those areas of their scalp that are going bald, meaning that these testosterone receptors are likely to have something to do with baldness. Following this observation, genetic researchers studying baldness focused their attention on the AR gene, which is responsible for producing the testosterone receptor. Variants within this gene were found to affect the number of testosterone receptors produced in the body and, together with other variants at different locations, have been associated with either protection against baldness or a significantly increased risk of going bald. Knowing whether you are at increased risk for baldness is important because the sooner you start using medications such as finasteride (Propecia) and minoxidil (Rogaine), the better the results, and knowing that you are predisposed to hair loss might encourage you to seek appropriate treatment sooner.
Altruism, musical ability, and dancing skill	AVPR1A, SLC6A4	Even complex social behaviors have a genetic component. The AVPR1A gene produces a protein that is active in specific areas of the brain involved with modulating behavior, and the SLC6A4 gene is involved in making sure that serotonin (which is a neurotransmitter) is circulated as needed throughout the brain. A combination of variants within these genes has been associated with various traits, including a predisposition toward altruism, musical ability, music appreciation, dancing skill, and even spirituality. In ancient times these traits and behaviors were extremely pertinent to our survival as a species. Communication through ritual movements, vocalization of wants and needs (especially between a mother and child),

continued

Altruism, musical ability, and dancing skill	AVPR1A, SLC6A4	courtship practices, and an ability to communicate and to cooperate within a group were extremely advantageous, and genes that enabled and promoted such behaviors helped our ancestors survive. Everything built into our genetic code today exists because it enabled our ancestors to survive and evolve over more than 5 million years. So, a genetic predisposition to musical ability and dancing is actually a remnant of yesteryear.
Chronotype	CLOCK, GNB3, PER1, PER2, PER3	The time of day when you are most productive is referred to as your chronotype; daytime people are referred to as larks, and evening people as owls. The genes involved in determining your chronotype produce proteins that are involved in the modulation of your internal clock, known as your circadian rhythm. And, yes, one of these genes has appropriately been named CLOCK. Variants within these genes have been shown to be involved in determining whether you are a morning or an evening person and have also been associated with sleeping patterns, such as an inability to go to sleep early or difficulty waking up in the morning. Because of their association with circadian rhythm, these genes have also been studied extensively in people with seasonal disorders. Some variants have been associated with becoming depressed during the winter, known as the winter blues or seasonal affective disorder. Fortunately, one can combat winter depression in a number of ways, such as exposure to artificial light that mimics the summer sunlight or by taking short courses of medications.

Amazing but true—many of the genetic variants we now find harmful to our health actually started out being beneficial to the survival of our distant ancestors centuries ago. Here's an example of how that can happen. For most of the time our species has existed, bleeding was a common cause of death. For women, bleeding after childbirth was directly associated with death, and profuse bleeding during a menstrual cycle contributed to iron deficiency, which, years ago, was also associated with a significantly increased risk of death. For men, bleeding after trauma, such as a hunting accident, was a feared cause of death. Then, about 24,000 years ago, a random variation occurred in a gene associated with the clotting of blood. The primary effect of this variant was that blood clotted much quicker, so people with the variant were much more likely to survive to adulthood, bear children, and pass on the beneficial variation to their offspring. It was truly Darwinian natural selection—people with this genetic variant were "selected" by nature because they were better fit to survive, and, therefore, the pro-clotting variant spread throughout the human population.

Now, however, we're living much longer than our ancestors did, and excessive bleeding is no longer life threatening in most developed countries. Not only that, but we are also doing things to our bodies that increase the risk of our blood clotting even when it isn't supposed to. For example, some of us smoke cigarettes, take medications such as birth control pills, use illicit drugs, are overweight, or lead a sedentary lifestyle, all of which increase our risk for blood clots. A genetic variant that increases clotting ability in combination with aging or any of these other lifestyle choices now significantly increases the risk of potentially harmful, or even deadly, blood clots, heart attacks, and strokes.

For literally *millions* of years, the average lifespan of our ancestors was no more than 20 to 30 years. However, in the past 200 years (and especially in the last 100), the average human lifespan has increased significantly and is now almost 80 years. Because human evolution occurs over tens of thousands of years, our genes have yet to catch up with our increased lifespan, and many of the new ways we are living today. In fact, because natural selection doesn't significantly affect genes that come into play after our reproductive years, there isn't much, if any, pressure on those genes to change. This means that the devastating diseases that affect us as we age will most likely *never* be filtered out by evolution and, there-fore, will continue to persist. It is up to us, then, to take over the responsibility for increasing our own ability to survive. We are capable of doing that right now by understanding and outsmarting those genes

Our species can no longer depend on evolutionary forces to have a significant effect on us. Going forward we must protect our health and wellness ourselves.

we have that are predisposing us to disease. The next evolutionary leap forward for our species lies in our ability to conduct genetic testing, predict risk of diseases based on the results, and prevent those diseases before they even manifest. Our species can no longer depend on evolutionary forces to have a significant effect on us. Going forward we must protect our health and wellness ourselves. We don't have to wait for future technology to do this—we have the ability today.

CONQUERING OUR SELFISH GENES

WE TEND to believe that our genes are just one component in the totality of who we are. But what if we looked at it a different way? What if our genes are not a part of us but, instead, we are a part of our genes?

In 1976, Richard Dawkins, a British evolutionary biologist, published a book called *The Selfish Gene*, in which he proposed a new theory of the relationship between a living organism, such as a human, and its genes. The theory states that genes, not humans, may be thought of as the true fundamental life-form. Rather than genes serving us, we exist to serve them, and the only reason we continue to exist is to ensure the survival not of ourselves but of our genes. Just take a moment to let the significance of this possibility sink in: We humans are mere hosts, and our genes are the true captains of the ship.

If Dawkins's theory is correct, and our genes' primary goal is to continue to be in existence for as long as possible, the only way for them to achieve this is by maximizing the reproduction of their hosts (us humans). Therefore, to a gene, a human is just a way for it to replicate and survive. When we have children, those children contain our genes, so when we eventually die, the genes live on—and so on through the generations. As long as our genes provide us with the ability to survive to childbearing age while also enabling our children to survive, our selfish genes will be passed down to future generations and, therefore, will have achieved their goal of immortality.

continued

And, as with most things, this concept also relates directly to sex. Why do humans and almost all animals have an innate drive to reproduce and protect their children as much as possible? One possible hypothesis is that it's because our genes, through their control of the neurochemistry in our brains, impart these innate needs. In fact, perhaps all our so-called innate needs are in some way gene induced.

Taking into account Dawkins's theories, it's interesting to speculate that our lifespan was so short for so many thousands of years because once we had fulfilled our childbearing requirements and helped raise our children to the age when they could take care of themselves, our genes no longer had any reason to keep us alive. Now, when we age, if our bodies get cancer or the arteries in our hearts become clogged or our brain gets dementia, our genes have little sympathy. They've gotten us to the point of successfully bearing and raising our children and anything after that is extraneous—extraneous to them but not to us.

Does this idea make you angry? It makes me very angry because I know that when I grow older—when I'm 60, 70, 80, and beyond—I'm going to want to live on, and I'm going to want my body and my mind to stay healthy. At that point, my genes may feel that my mission on this earth is complete, but I'm pretty sure I will not be in agreement.

In the past, we didn't have a choice. We were the hosts and our genes were in charge, dictating our destiny unchallenged. But today we can fight back by outsmarting our genes. We now understand the hierarchical relationship that has existed between us and our genes for more than 5 million years, and, through genetic technology, we can finally reverse that positioning. We can take the reins by taking control of our genes. The first step in this process is to know ourselves by figuring out exactly what genes and genetic variants we each contain.

This isn't to say that we can live without our genes—only that we can start to exert our control over the situation. We can significantly preserve and prolong our life by doing this—by understanding our genes and usurping their sovereign control.

0.5 PERCENT MAKES ALL THE DIFFERENCE

Genetically speaking, you are unique; there is no one else in the entire world with your exact genetic makeup. But it's also true that all humans are 99.5 percent genetically the same, which means that your genetic makeup is 99.5 percent the same as mine. So how can this be? How can we be unique and also 99.5 percent the same? The answer is that it's the 0.5 percent that makes all the difference.

Because your genetic makeup contains a total of 6 billion letters, variations in just 0.5 percent of that total means a difference of about 30 *million* letters, which is why we humans are so different from one another. Even though we may share 99.5 percent of the same exact genetic makeup, we still have about 30 million letters in our own genome that are different. Now, consider that changing just a single letter in your DNA can have a significant effect, such as giving you a different hair color, affecting your ability to taste various foods, determining your blood type, and giving you a particular disease. If each of these differences can result from changing a single letter, changes in 30 million different letters can clearly make quite a difference.

The real significance of the genetic code lies not in the percentage that makes us the same but in the variations that make each of us unique.

In terms of predictive medicine, the real significance of the genetic code lies not in the percentage that makes us the same but in the variations that make each of us unique.

THE HUMAN GENOME PROJECT

Once scientists understood how genes were passed down, how variations occurred, and how they determined virtually every aspect of who we are, it became clear that knowing the exact sequence of each one of those 20,000 genes and where exactly they appear on our chromosomes would be vitally important information.

Launched in 1990, the Human Genome Project was an international research effort whose result was a complete map of the exact sequence, end to end, of all the letters that compose the entire genetic makeup of a human. While the Human Genome Project was originally scheduled to be completed by 2005 at a total cost of $3 billion, Celera Genomics, a company run by Dr. J. Craig Venter, stated that they'd actually be able to do it quicker and for a fraction of the cost (only about $300 million). As always, competition created the need to innovate and become more efficient, which greatly accelerated the pace of the entire project. In 2000, a tie between the two (the international Human Genome Project and Celera Genomics) was announced by President Clinton, who credited both groups with the successful sequencing of the first full human genome. While the draft sequence was released to the world in 2000, the final sequence was actually completed in 2003.

Tens of thousands of genetic research studies are now being conducted around the world each year. The time has come to start moving all of this tremendously useful research from the laboratory into the doctor's office.

WHAT ALL OF THIS MEANS FOR YOU AND YOUR HEALTH

Since the completion of the Human Genome Project and tens of thousands of research studies, we are now able to test and analyze your entire genome and determine if your genetic makeup contains variants that are either harmful or beneficial. And we can then use that information to help you control and change your genetic destiny.

To perform the test, all we need to do is roll a swab around the inside of your mouth to gather a few cells. Cells for genetic testing can also be obtained from practically anywhere, including your saliva, blood, urine, skin, or hair. Because every cell in your body contains exactly the same genetic material, it doesn't matter where the cells come from. DNA can even be extracted from fingerprints themselves, so each time you touch something, such as a doorknob or a railing, you leave a bit of yourself behind.

It may seem odd that your skin cells, brain cells, and heart cells, for example, all contain exactly the same genes, but they do. The difference is that, through complex mechanisms, specific genes are turned on or off, depending on what type of cell it is. So even though your skin cells and brain cells contain exactly the same genes, only the genes specific to your skin are active in your skin cells and only those specific to your brain are active in your brain cells.

Genetic testing and the power of predictive medicine are available to you right now. The testing can be done on many different levels; we can sequence the genetic code of a single gene or we can test a large number of genes simultaneously to see if they contain any known variants. Cutting-edge technology, called full genome sequencing, now also allows us to sequence your *entire* genome, just like the researchers who completed the Human Genome Project.

FROM MENDEL TO PREDICTIVE MEDICINE: A TIMELINE	
The Past	
1866	Gregor Mendel's theories of inheritance are published.
1915	Thomas Morgan Hunt shows that genes, which exist on chromosomes, are responsible for inheritance.
1941	George Beadle and Edward Tatum show that genes control the functions of a cell because genes code for proteins.
1953	James Watson and Francis Crick discover the double helix structure of DNA.
1966	Marshall Nirenberg cracks the genetic code, which is the code of inheritance.
1978	Scientists genetically engineer bacteria to produce human insulin, which can be used to treat diabetes. This is the first drug made through the use of genetic engineering.
1983	Scientists identify the gene responsible for Huntington's disease, leading to one of the first genetic tests for a disease.
1990	The Human Genome Project is launched.
1994	Scientists locate the *BRCA1* gene and are able to start predicting a woman's genetic risk of breast cancer.

The Present	
2000	The Human Genome Project and Celera Genomics successfully complete an initial draft sequence of the entire human genome.
2006	The declining cost and increasing power of comprehensive genetic testing allow scientists to test for thousands of genes and their genetic variants all at a single time. This enables predictive medicine to become a reality and marks the beginning of medicine's pincer movement against disease.
2010	Full genome sequencing starts to become commercially viable, and individuals can now have their entire genome sequenced all at once. Full genome sequencing provides all the data required for full genome analysis, which, in turn, can provide information on virtually all known diseases and traits.
The Future	
~2018	Gene therapy is used to treat or potentially cure diseases such as cystic fibrosis and sickle-cell disease. More gene therapy treatments follow, including those for Parkinson's, Crohn's, and arthritis.
~2023	Genetic engineering research translates into medical care, allowing us to change specific genetic variants in a person, thereby changing disease risk on a genetic level.
~2028	Human longevity and wellness are significantly increased. Using genetic technology and predictive medicine, the average lifespan may be 125 years or longer (and these extra years will be active and full of vitality).

From the days of Mendel to the massive human genome sequencing endeavor of the Human Genome Project, the effect of every scientist's work throughout the history of genetics will play an enormous part in our fight to rid the world of disease. Keep turning the pages and we'll look at just how you can use genetic testing to live a healthier, longer life.

3

Genetic Screening
How It Works for You

MISCONCEPTION: Genetic testing and genetic analysis are really just two terms for the same thing.

FACT: Genetic testing and genetic analysis are actually two separate services. While each one is very complex on its own, successfully combining the two is the only way to get the most out of your genetic information.

I n the previous two chapters we discussed how and why your genes determine who you are mentally, emotionally, and physically. Throughout Part II, we'll discuss exactly how genetic testing, combined with predictive medicine, will allow you to outsmart your genes and live a longer, healthier life. Now I would like to explain just how genetic testing and genetic analysis are done so that when you choose to be tested, you will know exactly what to expect.

In the past, genetic testing and analysis were so expensive and time-consuming that they were primarily used to confirm the diagnosis of a rare disease that had already manifested. Now, however, advances in genetic technology allow us to conduct cost-effective, time-efficient genetic testing and analysis to *prevent* both rare and common diseases.

Genetic screening refers to any genetic testing and analysis that is conducted to evaluate your risk and carrier status for a number of diseases rather than for the purpose of definitively making or confirming a diagnosis. Therefore, most people who have genetic screening are healthy. Some, however, may also have an undiagnosed illness they aren't aware of because they don't yet have any symptoms, and some may already know they have a disease but want it analyzed comprehensively on the genetic level.

Genetic screening refers to any genetic testing and analysis that is conducted to evaluate your risk and carrier status for a number of diseases rather than for making or confirming a diagnosis.

GENETIC TESTING

The term *genetic testing* refers to the laboratory process that provides information about your DNA. While there are various kinds of testing, those that deduce the exact letters that make up your genetic code are the ones we'll be discussing here.

All that is needed to conduct a genetic test is a small number of cells taken from anywhere in your body. More than 25 years ago a chemist named Kary Mullis (who received the Nobel Prize for his discovery in 1993) developed a process called polymerase chain reaction (PCR), which makes it possible for scientists to make millions of copies of very small segments of DNA in a matter of hours. Because of this, scientists can now take a very small amount of your DNA from just a few cells and replicate it over and over again until they have all the DNA that's needed for testing.

Collecting the cells necessary for genetic testing is generally done in the office of a healthcare professional, although, because the process is as simple as rolling a swab around the inside of your mouth or even just spitting some saliva into a small container, some companies allow people to do it themselves at home. This means that genetic testing usually does *not* involve needles or blood. Once the cells are collected, they are sent to a laboratory where the DNA is isolated and the actual testing is conducted.

Genetic testing usually does *not* involve needles or blood.

Although there are many different laboratory technologies that can be used for genetic testing, traditional single-gene sequencing, arrays, and full genome sequencing are the methods used most often and are most relevant to you.

One way to conceptualize the differences among these three technologies is to think of genetic testing as watching a movie, with each letter of the genetic code representing a single frame and each gene representing one scene. Traditional single-gene sequencing shows us one *scene* at a sitting. Therefore, if we wanted to watch just that one scene this would be the best way to do it, but if we wanted to see the whole movie it would become very time-consuming since we're seeing only one scene at each sitting. Arrays, on the other hand, show us a large number of the *most important frames* taken from key scenes throughout the movie. Because we are able to view many different frames from a large number of scenes at once, we get a grasp of what the movie is about. But since we are able to view only a few frames from each scene, we're still missing a lot. Full genome sequencing shows every frame of the *entire movie* from the very beginning to the very end, in clear high definition. This method certainly provides the most information, but it's also the most expensive.

SINGLE-GENE SEQUENCING

The most conventional method of genetic testing focuses on the sequencing (that is, determining the complete genetic code) of just a single gene. Because your entire genetic makeup contains approximately 20,000 genes, this method provides only a very small fraction of the total. Single-gene sequencing can cost anywhere from a few hundred to a few thousand dollars.

- **Benefits:** Because the entire gene is sequenced, this approach is extremely thorough and will usually detect any and all genetic variations present in the gene. If you suspect that a rare disease may run in your family, this method is one of the most conclusive ways to test for it.
- **Pitfalls:** It is expensive, time-consuming, and inefficient to use when testing for more than a few genes at once. If you are concerned about a disease whose risk is determined by variants in many different genes, as most common diseases are, this is not a good method to use because it is too expensive and takes too long.

ARRAYS

Arrays go by many names, including genechips, microarrays, massarrays, and beadarrays. But whatever they're called, they all accomplish the same task, which is to determine the exact letters occurring at many different places within the genetic code all at a single time. Instead of deducing the sequence of an entire gene, arrays look at only specific letters in a number of genes. For the amount of information they provide, arrays are relatively inexpensive and usually cost just a few hundred dollars or less.

Although, on average, a gene contains about 6,000 letters, only five of those may be known to contain variants that are associated with disease. So, instead of looking at all 6,000 letters, arrays look at just those five, as well as at the most important letters in the next gene of interest, and so on. As can be seen, this process necessitates knowing exactly what we are looking for before we begin, because arrays detect only what they are designed to test for and skip over everything else.

Using this method, testing companies are able to look at anywhere from hundreds to millions of letters in your DNA at a single time. And because arrays can look at so many different letters, they can be used to gain important information on thousands of genes at once. While looking at hundreds or even millions of letters is still only a small percentage of the total, the specific letters about which arrays provide information are potentially the most important for your health; therefore, as long as the test is correctly configured to detect what matters most, an array actually provides tremendously useful information.

- **Benefits:** Arrays are inexpensive, have a fast turnaround time (from 8 to 72 hours), and have the unique ability to test for the most relevant genetic variants throughout thousands of genes simultaneously. This is an excellent way to test for common diseases that are caused by variants in many different genes because arrays allow you to connect all the dots.
- **Pitfalls:** Arrays are only moderately useful for rare diseases because they are not always configured to test for every single known genetic variant in a gene. Cystic fibrosis, for example, can be caused by more than 1,000 genetic variants *in a single gene*, but many arrays are designed to detect only the 5 or 10 variants that most frequently cause the disease; therefore, if your genetic code contains one of the rare variants not tested for, you may be told you don't carry a cystic fibrosis variant when you really do.

FULL GENOME SEQUENCING

Full genome sequencing is the pinnacle of genetic testing technology. While other testing technologies have provided us with pieces of information, this is the method by which we are able to determine *all 6 billion letters* of your entire genetic makeup *at a single time*. Thanks to the latest advances in genetic testing technology, what took the Human Genome Project 13 years and almost $3 billion to accomplish we can now do in a few days or weeks at a cost that is rapidly decreasing to the $1,000 mark. This means that the cost of full genome sequencing has decreased more than 99.99995 percent within the last 10 years, so just imagine what the next 10 years are going to bring.

When it was first introduced, the results produced by full genome sequencing were not as accurate as those obtained from single-gene sequencing or arrays; therefore, it was not suitable for use in medical decision making. However, the accuracy is quickly improving, and within a couple of years this technology will most likely be widely used by physicians working directly with patients.

- **Benefits:** Full genome sequencing ascertains your entire genetic code all at once, and it can be used to test for both rare and common diseases.
- **Pitfalls:** Because this technology is just now becoming available to the public, the accuracy needs to be perfected before it can be used in healthcare. However, the accuracy and usefulness of this new technology are improving at a rapid pace.

TECHNOLOGY	RARE DISEASES	COMMON DISEASES	NUMBER OF GENES TESTED FOR	TURN-AROUND TIME	APPROXI-MATE COST (IN DOLLARS)
Single-gene sequencing	Excellent	Poor	One	Weeks to months	Hundreds to thousands
Arrays	Moderate	Excellent	Thousands	Hours to days	Hundreds or less
Full genome sequencing	Excellent	Excellent	All	Days to weeks	Thousands

GENETIC ANALYSIS

While determining your exact genetic makeup through genetic testing is important, how that data are analyzed and presented to you are truly what makes the information invaluable. Genetic *testing* is a laboratory technique that deduces your exact genetic makeup; genetic *analysis* is the process of interpreting the meaning of that information, enabling you and your doctor to take action.

Let's say you've just had genetic testing and now you have a lot of data showing exactly what letters are contained in your genetic makeup. What do those letters mean? Do the results have any significance at all for your health? Does your genetic makeup show that you're at *increased risk* for, *carry*, or possibly even *have* a disease?

And, most important, what can you do, based on the results, to protect your health and wellness?

The way we arrive at the answers to each of these questions is by analyzing your genetic makeup to determine what it means and how you and your healthcare provider can use the information to outsmart your genes.

Genetic *testing* deduces your exact genetic makeup; genetic *analysis* is the process of interpreting that information so that it becomes useful for you and your doctor.

In less than a quarter century we have progressed from being able to conduct genetic testing for only one gene at a time to being able to perform a single test that identifies your entire genetic makeup. These newer testing technologies provide exponentially more genetic information. As a result, we are now able to conduct *comprehensive* analyses. Instead of being able to analyze your risk for just one disease, we can analyze your risk for hundreds at one time.

For the past several years, the majority of my own work has been focused on advancing the field of genetic analysis. The result is a new, more advanced way to analyze and convey genetic information so that it is as useful as possible to both you and your physician.

GENETIC SCREENING PANELS

Screening panels are a staple of medicine. You've probably heard one of the doctors on some television hospital drama shout, "I need a BMP, stat!" or Dr. Gregory House snidely commanding one of his residents to "Order LFTs and a CBC now, you buffoon!" All those letters are acronyms for screening panels. For example, CBC stands for complete blood count, a panel that includes a number of tests that measure different characteristics of a person's blood.

Panels allow for similar medical tests to be grouped together so that, instead of a doctor's having to know exactly what she's looking for in advance, she can let the results of the various tests provide her with insight into what's going on with the patient. A physician can be surprised to see that a value in one of the panel's tests is abnormal or different from what she had expected. Armed with this new knowledge, she is then better able to diagnose the patient correctly and provide the most appropriate treatment. Panels, therefore, improve a physician's diagnostic capabilities and increase her ability to practice good medicine efficiently.

Genetic screening panels are no different. They provide a way to organize a vast amount of information into defined categories that make the information manageable for both you and your physician. This is necessary because, with tests for so many diseases and traits now available, you need some way of determining which ones are important for you and which ones are not. Otherwise, you'd be overwhelmed and wouldn't even know where to start.

For example, if you are a woman you would probably choose a women's health panel, which is specifically formulated to screen for diseases and traits pertinent to you as a female. And for your children there's a children's panel, which tests for all of the diseases and traits pertinent to a child, including dyslexia, asthma, athletic predisposition to specific sports, risk of permanent hearing impairment if he or she listens to music too loudly, lactose intolerance, growth abnormalities, and preventable causes of sudden death.

Panels are extremely powerful tools because they enable simultaneous genetic screening for all the diseases and traits that are targeted to a specific need. In this way, the panel itself identifies what's going to be most important for you to focus on instead of you having to decide (or guess) in advance what to look for. For example, a 20-year-old man who had ordered a men's health panel might find that he is predisposed to having a heart attack even at a young age, but if he had been choosing from an à la carte menu of

> **Panels are extremely powerful tools because they enable simultaneous genetic screening for all the diseases and traits that are targeted to a specific need.**

diseases to test for, because of his age, heart attack would probably not have been on the list.

Panels are also a significant advancement in the field of genetic testing and analysis because, for the first time, screening for both rare and common diseases can be included in a single package. In the past, widespread screening for many of the rare diseases (often referred to as "orphan diseases" because of their rarity and the unfortunate lack of resources directed toward treating and preventing them) was discouraged because of costs and time. But now that genetic testing technology has become relatively inexpensive, we can start to include rare diseases in panels and begin to identify people who are carriers of or affected by them at the same time that we screen for common diseases. Through the use of genetic screening panels we can start to decrease the incidence of rare diseases while we concurrently battle the common ones, and orphan diseases will finally have found a home.

Through the use of genetic screening panels we can start to decrease the incidence of rare diseases while we concurrently battle the common ones.

Panels allow you to choose the group of diseases that are most relevant to you. Although many different panels exist, usually only one or two are specific to a particular individual, so choosing those that are most applicable is very straightforward and efficient.

The focus that panels provide not only makes it easier for you to choose the screening that is most appropriate for you and your family but also allows your healthcare provider to choose the one that is consistent with his or her specialty. For example, a cardiologist might want

Panels not only make it easier for you to choose the most appropriate screening but also allow your healthcare provider to choose the one that is consistent with his or her specialty.

information about your heart and blood vessels to determine whether you're at increased risk for a heart attack or for having an abnormal heart beat, but he would *not* want information on diseases that fall outside his field of specialization.

This kind of panel-based organization is becoming necessary because there are just so many possible individual diseases and traits we can now test for and analyze. Screening for just one disease at a time is quickly becoming antiquated and inefficient. However, screening for all known diseases provides so much information that the lack of focus significantly decreases its usefulness, and the most valuable information may end up being hidden in a pile of irrelevant results. Using panels solves both of these problems.

If you undergo genetic testing via arrays or full genome sequencing, the results will provide much more data than can be analyzed with just a single panel. And once you know your genetic makeup, it can be easily and confidentially stored and any part of it accessed and

Once you know your genetic makeup, it can be stored and any part of it accessed and analyzed at any time for the rest of your life.

analyzed at any time for the rest of your life. With panels, you get the relevant information when you need it, which makes the information much more applicable and, therefore, valuable to each stage of your life.

Here is a look at two genetic screening panels and what they analyze. Additional panels are presented in Part II.

Women's Health Panel

- Cancer, including breast, ovarian, endometrial, cervical, skin, and colorectal
- Heart disease, including heart attacks, blood pressure, coronary artery disease, and cholesterol levels

- Heart arrhythmias, including those that may cause sudden death
- Alzheimer's disease
- Stroke
- Multiple sclerosis
- Female infertility
- Polycystic ovary syndrome
- Premenstrual dysphoric disorder
- Migraine headaches
- Obesity and leanness
    - Body mass index (BMI), waist circumference, and fat accumulation in specific areas of the body (such as love handles)
- Genetically tailored nutrition
- Genetically tailored fitness, including exercise tolerance and predisposition to specific exercises and sports
- Diabetes, type 2
- Osteoporosis
- Caffeine metabolism, including whether caffeine is likely to affect sleep quality at night
- Risk of blood clots, including deep vein thrombosis (DVT)
- Depression, including the winter blues (seasonal affective disorder)
- Nicotine addiction, including effectiveness of nicotine cessation treatments
- Sensitivity to sunlight and tanning ability
- Arthritis, including osteoarthritis and rheumatoid arthritis
- Stomach ulcers
- Susceptibility to infectious diseases, including severity of and susceptibility or resistance to HIV/AIDS, SARS, West Nile virus, meningitis, hepatitis, stomach flu, and traveler's diarrhea
- Pharmacogenomics analysis, including effectiveness, adverse reactions, and dosing of medications pertinent to women

Men's Health Panel

- Cancer, including prostate, testicular, skin, and colorectal
- Heart disease, including heart attacks, blood pressure, coronary artery disease, and cholesterol levels
- Heart arrhythmias, including those that may cause sudden death
- Alzheimer's disease
- Stroke
- Male infertility
- Male-pattern baldness
- Erectile dysfunction (ED) medication treatment effectiveness
- Arthritis, including osteoarthritis and rheumatoid arthritis
- Obesity and leanness
  - Body mass index (BMI), waist circumference, and fat accumulation in specific areas of the body (such as love handles)
- Genetically tailored nutrition
- Genetically tailored fitness, including exercise tolerance and predisposition to specific exercises and sports
- Diabetes, type 2
- Arthritis, including osteoarthritis and rheumatoid arthritis
- Stomach ulcers
- Depression, including winter blues (seasonal affective disorder)
- Nicotine addiction, including effectiveness of nicotine cessation treatments
- Sensitivity to sunlight and tanning ability
- Caffeine metabolism, including whether caffeine is likely to affect sleep quality at night
- Risk of blood clots, including deep vein thrombosis (DVT)
- Susceptibility to infectious diseases, including severity of and susceptibility or resistance to HIV/AIDS, SARS, West Nile virus, meningitis, hepatitis, stomach flu, and traveler's diarrhea
- Pharmacogenomics analysis, including effectiveness, adverse reactions, and dosing of medications pertinent to men

THE DISEASE MATRIX AND REFLEX ANALYSIS

No disease exists in a vacuum, and all diseases are connected to other potential diseases and traits. A portion of your risk for a heart attack, for example, is determined by your genetic makeup, but your genetic makeup also determines your risk for many of the diseases and conditions related to a heart attack. Therefore, after finding out that someone has an increased risk for a heart attack, we can also provide an in-depth analysis of his or her risk for many of the diseases related to heart attacks as well as the preventions most likely to protect against those diseases. We can even analyze and provide information on the treatments that would be most effective if any of them should manifest. Based on all this information, the person will then be able to institute the most effective preventions and treatments to fight not only the primary disease but also all diseases that are related to it.

When a physician learns that you may be at risk of a heart attack, many questions arise, and genetic analysis can now provide useful information regarding each one:

- What is the *cause* of your increased risk for a heart attack?
    - Is it due to increased cholesterol levels, a blood clotting abnormality, or spasms of the blood vessels feeding your heart?
- What is the most effective way to *prevent* the heart attack?
    - Will exercise help and, if so, what's the best exercise regimen for you?
    - Will eating specific foods or drinking specific beverages reduce or increase your risk and, if so, which ones?
    - Will medications help and, if so, which will be the most and least effective medications?
    - Will specific lab values or biomarkers help in assessing risk?
- What is the most effective way to *treat* the heart attack, should it occur?
    - What procedures are most likely to be effective and which ones are most likely not to help?

- ▦ Which medications will be most and least effective?
- ● What related diseases are you also at increased risk for and what is your prognosis if you were to have a heart attack?
  - ▦ What is the chance of sudden death?
  - ▦ What is the chance of depression?
  - ▦ What is the chance of cognitive problems following certain procedures, such as bypass, that might be considered to avert or treat a heart attack?

The more answers we have, the more empowered we are to most effectively prevent and treat heart attacks on a personalized basis. And because the answer to each of these questions has been shown through research to be associated with our genes, comprehensive genetic analysis supplies information that helps to answer them all.

Let's walk through the disease matrix for heart attack so you'll see how it is constructed and how valuable it is for predicting, preventing, and treating a disease. Each disease or trait (for simplicity, we'll refer to them all as *diseases*) is represented by an oval, and the interconnectedness between two diseases is represented by a solid line. The shading of the ovals represents the various levels of the disease matrix. The black oval is the primary disease (in this example, it's a heart attack); gray ovals represent secondary diseases that are directly associated with the primary one; polka-dotted ovals are tertiary diseases that are directly associated with the secondary diseases; and so on. The matrix can contain as many levels as necessary to identify all possible diseases related in any way to the primary disease.

We start with the primary disease of interest, heart attack, which is at the center of the heart attack disease matrix.

Then we connect a disease that's directly related to heart attack, such as coronary artery disease.

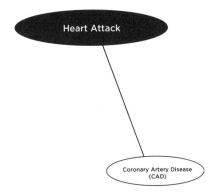

We continue on to connect all other secondary diseases, traits, preventions, or treatments that are directly related to heart attack.

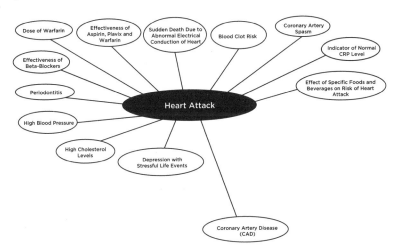

Then we go to the third level—diseases, preventions, or treatments directly related to any of the second-level diseases. For simplicity, let's consider only the two that are attached to coronary artery disease. Analyzing your genetic code further will provide actionable information on treatments of that disease, including whether a medication is going to be effective for lowering your cholesterol levels and whether you are likely to have an adverse reaction to a statin, which is the most commonly prescribed class of cholesterol-lowering medications (such as Lipitor, Zocor, and Crestor).

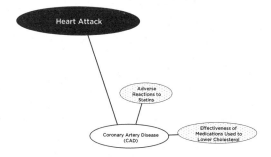

Each of these steps is repeated until the entire disease matrix is constructed. The complete matrix for heart attack will look like the one shown below. While it may look a bit overwhelming on a printed page, it is actually much easier to read and navigate in its native format, which is a three-dimensional, fully interactive, computer-generated color model displayed on a large computer monitor.

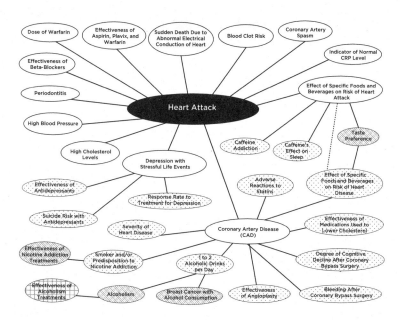

Each disease, rare or common, has its own matrix, and those two classifications of diseases aren't mutually exclusive because many times one or more rare disease exists somewhere within the matrix for a common disease, and vice versa. For example, in the disease matrix for heart attack, high cholesterol includes all possible causes of high cholesterol, from the rare forms (such as familial hypercholesterolemia) to the more common form with which most people are familiar. The comprehensiveness of the screening, therefore, includes not only a large number of diseases but also all possible causes of each of those diseases. Additional examples of disease matrices can be found at www.OutsmartYourGenes.com/DiseaseMatrix.

The next question, of course, is how to make information provided by the disease matrix actionable. The answer is by using another technology called reflex analysis. Here's how it works. If, for example, you or your physician had ordered a woman's health panel, and you were found to be at increased risk for a heart attack, a specially designed computer program would perform a reflex analysis, which would automatically analyze your particular risk for all the secondary diseases (the gray ovals) in the disease matrix for heart attack. If you were found to be at increased risk for any of the secondary diseases, the computer program would then analyze all of the connected third-level diseases (the polka-dotted ovals), and so forth, until all of the applicable diseases had been analyzed. This type of fully integrated, comprehensive genetic analysis is now possible because arrays and full genome sequencing provide us with a wealth of information about thousands of genes at once.

The disease matrix defines the interconnectedness of diseases for which you are found to be at risk, and reflex analysis makes the results of your genetic analysis personalized, integrated, and highly actionable.

Panels are a way to intelligently organize all relevant diseases and traits for genetic screening; the disease matrix then defines the interconnectedness of diseases and traits for which you are found to be at risk, and reflex analysis is the process by which we are able to use the disease matrix to make the results of your genetic analysis personalized, integrated, and highly actionable.

The integration of reflex analysis into the analytical process provides an exponentially greater amount of information and significantly increases its value to you and your physician by laying out a prevention and treatment plan *genetically tailored* just for you. But if it isn't organized well, the sheer volume of information provided may become too overwhelming, and that's where the next-generation genetic report comes in. A modern, well-laid-out genetic report presents the information in a format that is clear, straightforward, and accessible, which makes it that much easier to act upon.

THE GENETIC REPORT

After genetic testing and analysis, all of the results are usually compiled into a genetic report. This genetic report shouldn't leave you or your doctor wondering, "What the heck does that mean?" A well-written, concise, and uncomplicated report allows you both to efficiently process all the information it contains, with a specific focus on making *all* of its information actionable. Otherwise, no matter how advanced the genetic testing or how complete the analysis, the results won't do you any practical good.

> **A well-written, concise, and uncomplicated genetic report allows you to efficiently process all the information it contains and focuses on making all of its information actionable.**

A genetic report does not need to contain any difficult terminology whatsoever. I've seen reports using words like *genotypes*, *alleles*, *odds ratios*, *SNPs*, and *polymorphisms*, because there is a school of thought that believes everyone ought to have some knowledge of the science and terminology of genetics before receiving genetic information. However, my approach is to write the report so that anyone, regardless of his or her expertise in the field of genetics or medicine, can understand it.

The job of your physician is to assume that you do not know anything about her specialty and, therefore, to convey all information in terms that are as simple and straightforward as possible. Predictive medicine is no different. If you have a desire to read about the science of genetics, there are many resources available, from popular websites to scientific texts, but a well-written genetic report should never assume that you've read any of that material or require you to know anything about genetics.

A sample page from a report on the results of a women's health panel appears on the next page. I have only included the heart attack results, but the overall report includes many other disease results. To see more pages from a sample report, visit my website at www.Out smartYourGenes.com/Report. This reporting format was invented after talking with many people who had received genetic reports filled with convoluted and tedious information that seemed to assume the recipient had a PhD in the field of genetics.

As these genetic report pages show, information about each disease, including a concise description, the results of the genetic testing displayed as easy-to-understand risk values, and actionable preventive measures, are organized and straightforward. The report also includes graphic representations of other important values, such as the disease's clinical significance (how seriously it may affect your health and wellness) and if there are actions available to significantly reduce the risk of the disease.

Because physicians are often pressed for time and want to access all pertinent information as quickly as possible, the portion of the report directed specifically to the healthcare provider summarizes all the key information in two to three pages and presents the findings most relevant to clinical care first. There may be a bit more genetic terminology used in this portion of the report, and it may also contain physician-specific information, such as dosing suggestions for specific medications, that is not included in the primary report. Making it as easy and as efficient as possible for both you and your doctor to understand genetic information encourages your doctor to embrace genetic screening as a way to provide you with truly personalized medicine.

GENETIC REPORT

CIN 0042882220

Heart Attack

A heart attack occurs when the blood supply to the heart is interrupted, causing some heart tissue to die. A heart attack is an emergency because it may cause sudden death or, following a heart attack, the heart may not function correctly. The most common cause is a build-up of plaque in one of the blood vessels feeding the heart, a condition referred to as coronary artery disease, which is related to high cholesterol levels.

Over one million people suffer from heart attack each year in the United States and about 40% of these people will die from the heart attack. Heart disease, including heart attacks, is the leading cause of death in the world but numerous preventive measures exist.

Your Predictive Medicine Lifetime Risk

61%

Risk

Generic Lifetime Risk = 32%
Your Lifetime Risk = 61%
(This is equal to a 90% increased risk)

Clinical Significance

This potential disease is
<u>very important</u>
to your health and wellness.

Actionability

Preventive measures have
been shown to help prevent
or slow down this disease.

Onset & Symptoms

You are at greatest risk of a heart attack after the age of 40. Symptoms of a heart attack includes chest pain that may radiate down the left arm or up to the jaw as well as shortness of breath, fatigue, sweating and nausea and/or vomiting. A heart attack is an emergency situation and if you think you are having a heart attack, you should call 911 or immediately be brought the nearest emergency room.

Genetically Tailored Prevention

Monitoring

❖ Please discuss your risk and the following preventive measures with your doctor or a cardiologist.

❖ You are also predisposed to *coronary artery disease, high cholesterol levels* and *high blood pressure*, each of which is associated with an increased risk of a heart attack. Be vigilant of each of these as controlling them may significantly reduce your risk of a heart attack.

❖ Numerous tests can be administered by your doctor, including blood tests and radiologic exams, to detect and monitor for high cholesterol levels, coronary artery disease and heart attacks.

Lifestyle Modifications

❖ Drinking one 5-ounce glass of red wine per day may decrease your risk of high cholesterol levels and protect your heart against injury. You are not predisposed to alcoholism but consuming *more than* one alcoholic drink per day may have negative consequences on your health.

❖ Eating at least one cup per week of broccoli, collard greens, cabbage, kale, Brussels sprouts, or bok choy, may decrease your risk of a heart attack. However, you may not like the taste of these vegetables so consider mixing them with other food to make them more palatable.§

❖ Your heart attack risk is genetically linked with an increased risk of periodontitis. Good oral hygiene & annual dental exams may decrease your risk of both diseases.

Medications

Effective
❖ You are sensitive to the blood-thinning medication Warfarin (Coumadin®) and a lower dose may be required in-order to avoid adverse reactions, such as bleeding.

❖ Beta-blockers and aspirin will most likely be effective.

❖ Statins should be effective in lowering your risk of cardiovascular disease. You are not predisposed to adverse reactions with statins.

Not Effective
❖ Clopidogrel (Plavix®) may be ineffective in reducing your risk of a heart attack and cardiovascular disease.

Disease Interventions

❖ Angioplasty may be effective in unclogging your arteries and you have a lower than average chance of your arteries reclogging quickly after this procedure.§

❖ If coronary bypass is required, you are not at an increased risk of cognitive decline due to this procedure.§

❖ You don't have an increased risk of depression following a heart attack but if you do experience depression, antidepressant medications (SSRIs) should be effective.

Additional Information

❖ American Heart Association: www.americanheart.org

§ = Preliminary association that requires the support from additional research studies before it can be considered definitive information.

GENETIC REPORT
CIN 0042882220   *Confidential Information*

Colorectal Cancer

Colorectal cancer is cancer that forms in the colon or rectum. It is the third most common cause of cancer and also the third cause of all cancer related deaths.

There are around 150,000 new cases of colorectal cancer diagnosed each year. However, the American Cancer Society and most physicians believe that death from cancer can almost always be prevented as long as it is detected in its early stages. There are also numerous preventive measures that have been shown to decrease the risk of colorectal cancer and as well as numerous screening modalities to ensure that it is detected at its earliest possible stage should it ever occur.

Your Predictive Medicine
Lifetime Risk

21%

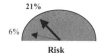

Risk
Generic Lifetime Risk = 6%
Your Lifetime Risk = 21%
(This is equal to a 250% increased risk)

Clinical Significance
This potential disease is
<u>very important</u>
to your health and wellness.

Actionability
Preventive measures have
been shown to help prevent
or slow down this disease.

Onset & Symptoms

Based on your genetic code, this disease, should it manifest, will most likely affect you after the age of 40. Oftentimes symptoms will not become apparent until *after* the cancer has grown to a significant size. Because of this, regular colonoscopy screening, which has the power to detect this cancer at an early and treatable stage, can be life saving. Colorectal cancer may cause bleeding into the colon that may be visually apparent as blood in the stool. Other symptoms may include unexplained weight loss and/or fatigue.

Genetically Tailored Prevention

Monitoring
❖ Please discuss your risk and the following preventive measures with your doctor or a gastroenterologist.
❖ Due to increased risk, a traditional colonoscopy (not a virtual one) every five years starting at age 40 is indicated in-order to appropriately screen for precancerous changes, such as polyps, and cancer.
❖ A Fecal Occult Blood Test (FOBT) once a year during the four years in-between each colonoscopy may provide additional safeguard.
❖ Report any significant weight loss for an unknown reason or blood in stool to your physician immediately.
❖ Check vitamin D levels with a blood test each year to make sure levels are normal. If low, discuss taking vitamin D supplements with your doctor.

Lifestyle Modifications
❖ Decrease consumption of red meat to no more than one serving per week as your genetic code significantly increases your risk of colorectal cancer if red meat is consumed on a more frequent basis.
❖ Avoid eating any processed meats such as hot dogs, sausages, salami, bacon, ham, pastrami, and cold cuts, and any other cured meats.
❖ Proactively avoid exposure to second hand smoke. Based on your genetic code, you are at significantly increased risk of colorectal cancer due to repeated exposure to second hand smoke.§
❖ Avoid smoking cigarettes or other tobacco products.
❖ Increase physical exercise. The more you exercise on a regular basis each week, the lower your risk.
❖ Avoid becoming overweight or obese, especially in-regards to abdominal fat. You are not predisposed to being overweight or obesity.

Medications
Over-the-counter Medications:
❖ Aspirin taken once a day only if approved by your doctor.
❖ Vitamin D 1,000 IU once a day.

Complementary Interventions
❖ Eat one or more cloves of fresh garlic per day.§

Misconceptions
❖ Vitamin E supplements have been shown to *not* be effective in reducing colorectal cancer risk and studies have also shown it may be harmful to overall health.

Additional Information
❖ Colon Cancer Alliance: www.ccalliance.org
❖ American Cancer Society: www.cancer.org

§ = Preliminary association that requires the support from additional research studies before it can be considered definitive information.

GENETIC REPORT
CIN 0042882220

Confidential Information

Healthcare Provider Summary

Emergency Alert	Increased Risk	Carrier	Decreased Risk
Malignant Hyperthermia	Myocardial Infarction	Cystic Fibrosis	Macular Degeneration
	Colorectal Cancer	G6PD Deficiency	Melanoma
	Alzheimer's Disease		Multiple Sclerosis

Increased Risk

❖ Increased risk for **Myocardial Infarction**

- o Based on a comprehensive analysis of their genetic profile, this person has a **61%** lifetime risk of a MI after the age of 40, compared to the 32% risk for the general population. Prevention may be effective in decreasing their risk.
- o This person is also at increased risk of *CAD, high LDL levels,* and *HTN.*
- o Relevant Preventive Measures: (decreases incidence and/or slows progression of CAD and MI)
  - *Screening:*
    - ➢ Monitor blood pressure and cholesterol levels on a regular basis.
    - ➢ Consider stress echos or similar tests to rule out the existence of CAD.
    - ➢ Check CRP levels annually and consider initiating statin if CRP is elevated. According to their genetic makeup, in a non-disease state this person's CRP levels should fall within the normal standard range.
  - *Medications:*
    - ➢ **Warfarin Sensitivity** – consider lower starting dose to avoid bleeding. Studies indicate that a dose of 2.9 mg/day may be effective and safe, and should reach a therapeutic INR around day 7.
    - ➢ Clopidogrel may be ineffective as an antiplatlet agent in this person and may not protect against MIs.
    - ➢ Beta-blockers, Aspirin and statins should all have normal effectiveness.
    - ➢ No increased risk of myopathy with statins.
  - *Lifestyle & Alternative Therapies:*
    - ➢ Eating ≥ 1-2 servings of cruciferous vegetables per week may decrease their risk of a MI.§
    - ➢ Drinking 1 glass/day of red wine will help increase HDL, decrease LDL, and precondition heart to ischemia.
    - ➢ This person's increased risk of MI is genetically linked to an increased risk of periodontitis. Consider checking teeth & reinforcing good dental hygiene at each visit. Encourage regular check-ups with a dentist.
- o Relevant *Treatment and Intervention* information if CAD or an MI should ever occur:
    - ➢ Coronary Angioplasty - lower than average risk of restenosis.§
    - ➢ CABG - no increased risk of cognitive decline.§
    - ➢ Depression – no increased risk of depression with stressful life events.
      - • SSRIs should have normal effectiveness in treating depression or depressive symptoms.

❖ Increased risk for **Colorectal Cancer**

- o Based on a comprehensive analysis of their genetic profile, this person has a **21%** lifetime risk of CRC, compared to the 6% lifetime risk for the general Caucasian female population. Onset most likely to occur after the age of 40.
- o Relevant Preventive Measures: (decreases incidence and/or may augment chances of detecting CRC at earliest stage)
  - *Screening:*
    - ➢ Due to increased risk, a traditional colonoscopy (not a virtual one) every five years starting at age 40 is indicated in-order to appropriately screen for precancerous changes, such as polyps, and cancer.
    - ➢ FOBT once a year during the four years in-between each colonoscopy may provide additional safeguard.
    - ➢ Check vitamin D levels with a blood test each year to make sure levels are normal.
  - *Medications:*
    - ➢ Aspirin may provide preventive benefit against CRC, MI, and Alzheimer's. Consider concurrent PPI.
    - ➢ Vitamin D3 supplementation, such as 1,000IU QD.
  - *Lifestyle & Alternative Therapies:*
    - ➢ Person is genetically predisposed to colorectal cancer if they consume red meat more than once per week.
    - ➢ This person indicated they are not a smoker nor are they exposed to second-hand smoke on a regular basis. Consider checking with patient annually to reconfirm that they are not smoking as their genetic code significantly increases their risk of cancer with smoking.§
    - ➢ Long-term physical exercise will decrease the person's risk of MI, Alzheimer's, and CRC.
    - ➢ Monitor BMI annually as being overweight or obese appears to increase the risk of CRC.
    - ➢ Increased consumption of garlic may lower their risk of CRC.
- o Relevant *Treatment and Intervention* information if CRC should ever occur:
    - ▪ 5-FU – Normal response and no increased risk of toxicity.
    - ▪ Irinotecan – Increased risk of neutropenia and diarrhea at higher doses due to UGT1A1*28/*28.
    Absolute risk of toxicity at lower doses (100–125 mg/m2) = 15%
    Absolute risk of toxicity at intermediate doses (150–250 mg/m2) = 25%–40%
    Absolute risk of toxicity at higher doses (>250 mg/m2) = 50%–70%
    For reference, if this person did not possess UGT1A1*28/*28 then their risk of toxicity at any dose would be 15%.

§ = Preliminary association that has not yet been replicated through additional studies or within this person's specific ethnicity or gender.
Please contact us at (XXX) XXX-XXXX with any questions or if you require clarification on any of the information contained in this report. Prepared by Brandon Colby, MD.

4

Become Informed About Genetic Screening
Who, What, and Where

MISCONCEPTION: All genetic testing and analysis companies are basically the same, so I might as well use the one that's least expensive.

FACT: Caveat emptor! This is an industry like any other, and it's not very well regulated, so there is actually substantial variability among the companies and laboratories that provide genetic testing and analysis. Understanding which questions to ask and where to look for answers will enable you to make an informed decision about what is best for you and your family.

Genetic testing and analysis (which I'll now refer to solely as genetic screening) is primarily provided by public and private companies. In the past few years, companies from many different industries have moved into the genetic screening market. They include companies founded and run by licensed physicians, companies composed primarily of PhDs untrained in the practice of medicine, laboratories, venture capitalists, and even Internet-based dot-coms. Their services may be delivered either through a physician or direct to consumers over the Internet.

Because there are so many options, and because not all companies' products and services are equally useful, it is important that you become an educated consumer *before* proceeding with genetic screening.

In this chapter I'll provide an overview of the genetic screening process and the personal genomics industry, and then I'll discuss key issues you'll want to consider, such as confidentiality and industry regulation. Lastly I'll tell you the five most important questions to ask before you have genetic screening (and the answers you'll want to receive).

BEHIND THE SCENES OF GENETIC RESEARCH AND DEVELOPMENT

The standard genetic screening company does four things:

1. Uses the information gathered by genetic researchers to formulate and keep its services up-to-date
2. Provides genetic testing
3. Provides genetic analysis
4. Delivers the results

The way in which these services are performed is what differentiates one company from another; very few of them do all four equally well. For example, a company may do great research and use a reliable laboratory for testing but may fall short on genetic analysis and the way they deliver the results.

RESEARCH GENOMICS VS. PERSONAL GENOMICS

Although the word *genetics* is familiar to most people, many haven't yet seen the word *genomics*. Genetics primarily means the study of genes, and genomics refers to the study of all genes and how those genes interact with nongenetic factors.

To understand the genomics industry, it's important for us to briefly discuss the distinction between two related fields: research genomics and personal genomics. Research genomics takes place at an academic institution, a hospital, or another type of research organization and focuses on trying to decipher the exact genetic cause of a disease. If you've ever seen a newspaper headline that read, "A Gene for Disease *X* Has Been Found," the article is referencing the work of research genomics. This type of research has been ongoing for more than three decades. At first it was focused primarily on the rare diseases because they are generally caused by only one or two genetic variants within a single gene and are, therefore, relatively straightforward. In the past decade, however, researchers have been able to collect an enormous amount of genetic data by using arrays, and this has enabled them to identify a significant number of the variants involved in predisposing people to almost all known common diseases.

After the researchers have completed and published their studies, the next step is putting the results of their research into practical use for the individual—that is, moving from research genomics to personal genomics. While research genomics uncovers the specific genetic variants associated with disease, personal genomics uses that information to provide genetic screening for the individual. Any company or institution that conducts genetic screening on an individual and provides results to a person or a person's physician is working in the field of personal genomics.

While research genomics uncovers the specific genetic variants associated with disease, personal genomics uses that information to provide genetic screening for the individual.

Predictive medicine is at the cutting edge of personal genomics because it makes genetic information actionable. Whereas personal genomics provides you with access to genetic screening, predictive medicine tells you what you can do about it.

MOVING FROM RESEARCH TO PERSONAL GENOMICS

Unlike the scientists working in the field of research genomics, companies involved in personal genomics do not generally conduct studies of large groups of people. Rather, they use the information published by researchers, which is available to everyone.

The type of information provided by personal genomics companies results from how each individual company decides to use the research studies available to them. Those decisions may involve determining whether to use single-gene sequencing, arrays, or full-genome sequencing; whether to conduct the genetic testing in their own lab or to outsource it; which specific genetic variants to use when analyzing the risk for a particular disease; how to analyze the results; and how to deliver the results.

Some companies screen for only a small group of diseases, whereas others have the ability to screen for virtually all known diseases. Some companies screen for only rare diseases, other companies screen primarily for common diseases, and a few screen for both. However, the differences go beyond disease categories and apply also to the specific variants the company decides to test for and analyze in relation to each disease. For example, some companies use only a few of the most common variants, whereas others incorporate a much larger number.

Clearly, looking at a large number of both rare and common diseases and at a large number of variants associated with each of those diseases provides the most information, which, in turn, will allow for the most comprehensive genetic analysis. There are, however, only a few companies that undertake this type of in-depth, comprehensive approach.

Looking at a large number of rare and common diseases and at a large number of variants associated with each of those diseases provides the most information and allows for the most comprehensive genetic analysis.

Once the decision is made as to what diseases and genetic variants to screen for and what testing technology to use, the company must then focus on choosing a laboratory to conduct the actual genetic testing.

THE LABORATORY

Without genetic testing, there's no data for genetic analysis, and because genetic testing can currently be performed only in a laboratory, all personal genomics companies are associated with one or more labs. Therefore, when you choose a personal genomics company, you're also choosing their laboratory, which means that the laboratory is an important consideration that should not be overlooked.

Some personal genomics companies own their own laboratory, and others contract out the lab work. Although it doesn't really matter to you whether the company owns a lab (because it has no effect on the accuracy or quality of the laboratory work), it is very important for you to ensure that whatever laboratory the company uses has either CLIA or ISO-15189 certification (see page 81), which means it is regulated and certified by an independent federal agency. The most important thing for you to understand right now is that there is tremendous variability in the laboratories that offer genetic testing and that not all of them are regulated.

When you choose a personal genomics company, you're also choosing their laboratory.

GENETIC ANALYSIS

After the testing is completed by the lab, the next task for the personal genomics company is to conduct the genetic analysis. As discussed, there is a significant difference between simply providing you with results of the testing and actually conducting in-depth analysis to make the results truly actionable.

RESULTS DELIVERY

The final responsibility of a personal genomics company is to provide a report of the results. Some companies provide genetic screening services only through healthcare professionals, such as doctors, who then offer these services to their patients. Other companies provide their services direct to the consumer and thus circumvent the physician. Although almost all of the direct-to-consumer companies are dot-coms that provide their results via the Internet (you log on to their website to view your results), the majority of companies that provide genetic screening only through healthcare professionals supply their results in a printed report.

In terms of the genetic report itself, the three differentiating factors are actionability, integration, and clarity. Some companies, for example, provide actionable genetic reports that can be used to outsmart your genes, while others shy away from providing you with this type of information (for example, they may tell you have an increased risk of Alzheimer's disease, but they don't include any information about what that means or how to decrease that risk).

Some companies may also provide fully integrated results by using panels, disease matrices, and reflex analysis; others provide results that are not integrated at all. Some may provide results that are extremely convoluted and primarily written for a geneticist, and other companies provide more concise, straightforward reports that can be read and understood by all. A list of personal genomics companies appears at the end of this chapter.

AN OVERVIEW OF THE ENTIRE PROCESS

Here's a closer look at the entire genetic screening process. After some cells from your body are collected and delivered to a personal genomics company, the company sends the cells to a laboratory. The laboratory is able to extract and purify the DNA that's contained within those cells. Genetic testing is then performed on the DNA using single-gene sequencing, arrays, or full genome sequencing, and the result is a string of letters representing your genetic makeup. The amount of genetic information can be very small, such as one or two letters from a single gene, or it can be gargantuan, such as all 6 billion letters of your entire genetic makeup. This information is then sent electronically from the laboratory to the personal genomics company. The personal genomics company performs genetic analysis on the information, creates a report, and delivers the results either to your healthcare professional or directly to you. This entire process, from beginning to end, can take anywhere from a week to a few months, depending on the efficiency of the laboratory and the personal genomics company.

The rest of this chapter provides you with the information you need to make informed decisions about genetic screening. If you know what to watch out for when evaluating a personal genomics company, the results you receive are more likely to be useful in benefiting your health.

DIRECT-TO-CONSUMER GENETIC SCREENING

Direct-to-consumer genetic screening refers to testing and analysis that is provided directly to you without your having to go to a physician or other healthcare professional in any way. The direct-to-consumer genetics market is actually quite new, and because it circumvents the medical establishment, some countries (such as Germany) have banned it. Of the many companies providing this type of service, some are medically focused, and others state that their genetic testing is only for novelty, entertainment, recreational, or informational purposes, meaning that the results they provide are for informational purposes.

GENEALOGICAL (ANCESTRY) GENETIC TESTING

GENEALOGICAL GENETIC testing, which is conducted to determine your ancestral and ethnic origins, is one of the forms of testing that is, for the most part, done for its novelty value without a direct medical application. In essence, it's like building a genetic family tree.

While genetic testing can't tell you specifically who your great-great-great-grandfather was, it can determine that your ancestors were most likely Native American, European, African, Asian, East Asian, or Cohanim (that is, a direct descendant of the biblical Aaron, brother of Moses). Some services even show you the migration your distant ancestors most likely took from their original location to the present day, along with when that migration most likely occurred. Others help you build a family tree by finding other individuals who had genealogical genetic testing done by the same company and who are related to you. (The tree may have very few branches, however, depending on the number of your relatives who have signed up for the service.)

The majority of genealogical genetic testing services are Internet-based companies that charge anywhere from under $100 to several hundred dollars. You go to the company's website, pay online, receive a test kit in the mail, submit your DNA sample, and after a few weeks you'll receive your results on the company's website or via email.

The only time this kind of testing may be medically useful is if a person is adopted or unsure of his or her ethnic background. Because some genetic variants are associated with increased disease risk only in a specific ethnicity, it can be important to know such information. For example, a genetic variant that increases the risk of multiple sclerosis in Caucasians may not increase the risk in Japanese.

As with all genetic testing, you should check to make sure the laboratory used is CLIA-certified (see page 81). If it's not, there won't be any governmental oversight and the quality of the results you receive may be questionable. Because you want to make sure that you'll understand the results, it is always a good idea to look at a sample report *before* you commit. If there is no sample report accessible on the company's website, you should be able to request one via email. A reputable company will be happy to comply.

As a physician, the primary issue I have with direct-to-consumer testing is that, no matter what they claim, these companies *are* providing information that is inherently and indisputably medically relevant. Some, for example, determine your risk of developing breast and colon cancer, multiple sclerosis, Alzheimer's disease, and Parkinson's disease, yet they claim that the information is not medical in nature. Even though the testing and analysis many of these companies perform is very advanced and usually quite reputable, I do believe that because of its medical relevancy a healthcare professional should always be involved in the process.

"Direct-to-Consumer Genetic Tests: Flawed and Unethical," a 2008 commentary published in *Lancet Oncology*, notes that having a healthcare professional involved in genetic screening for disease risk assessment is essential but is often not provided by companies that claim the information they provide is for recreational purposes only.

A small study conducted in 2008 by researchers at the University of Chicago and the Medical College of Wisconsin found that nearly 100 percent of physicians believe that genetic screening requires counseling by a healthcare professional, and more than 90 percent believe that the sale of genetic screening directly to the consumer should be restricted. In the United States, however, the government has yet to take a legal stance on direct-to-consumer sales, the minimum services a genetic screening company must supply, or even the terminology a company can use when marketing its services. The

Council of Europe, on the other hand, has proposed legislation to its 47 member nations that would make genetic screening for any medical condition permissible only through a physician.

OTHER PROVIDERS OF GENETIC SCREENING

One growing trend in healthcare is the establishment of clinics that use various assessment techniques to provide personalized preventive medical services intended to help people avoid illness, slow down aging, increase longevity, and maintain vitality for as long as possible. While some of these clinics currently offer some degree of genetic screening, I believe that within the next few years the majority of them will be integrating comprehensive genetic screening and predictive medicine into their services. As an alternative to your own healthcare provider or direct via the Internet, these specialized clinics represent another way for you to receive these services.

FEAR, ANXIETY, AND GENETIC SCREENING

The most pressing question most people have when they consider genetic screening is, "Do I really want to know if I am predisposed to a disease? Will the anxiety from knowing be so great as to diminish any possible benefit?" This is a normal fear of the unknown, and over the past few years various studies have evaluated anxiety levels before and after genetic screening to determine if the information is doing more harm than good. A moderate amount of anxiety in the short term may be beneficial because it increases awareness of a potential danger and inspires the person to take action, but severe anxiety over many months or years could potentially cause persistent negative emotions or depression. So, what have these studies shown?

In 1999, researchers at King's College London and the University of Leicester in the United Kingdom conducted an extensive review of more than 50 studies evaluating the psychological impact of predicting a person's risks of such illness as heart disease, cancer, osteoporosis, diabetes, and neurological diseases. They assessed a number of factors, including anxiety, distress, depression, perceived well-being, and even work absenteeism. What the researchers found is that there was increased psychological distress *only* in the short

Counseling by a healthcare professional successfully reduces psychological distress.

term (for less than 1 month after receiving the results) but *not in the long term* (after 1 month). They also found that the potential for psychological distress can be successfully mitigated with counseling by healthcare professionals.

Since some degree of anxiety and distress are to be expected when you have any kind of medical test that provides information about a disease, these findings aren't surprising. The key, however, is that counseling by a healthcare professional does successfully reduce any psychological distress that might occur.

A more recent review of the literature conducted in 2008 and published in *Genetics in Medicine* evaluated 30 studies assessing the psychological impact of genetic screening for colon cancer, breast cancer, ovarian cancer, and Alzheimer's disease on the individual. The article concluded that genetic testing did not have a significant negative psychological effect. Other studies have found that individuals who learned they were predisposed to a disease were more likely to engage in long-term preventive measures that would lower their risk.

But what if your results indicate that you do *not* have an increased risk or that you actually have a *lower risk* of a disease? Some healthcare providers fear that, in this situation, genetic screening may actually lead to a false sense of security. For example, if a person with a family history of melanoma undergoes genetic screening and is found *not* to be predisposed to the cancer, would that person be less diligent about wearing sunscreen?

A small study conducted in 2009 by researchers at the University of Utah found that this was not, in fact, the case. Individuals who learned that they did not have an increased risk of melanoma still continued to protect themselves from the sun. Furthermore, while these individuals started to spend more time outdoors, they also started to wear sun-protective clothing. One possible explanation is that all the study participants received counseling from a healthcare provider so that they understood exactly what the results of the genetic screening meant and were told that protecting themselves from the sun was still important.

These findings complement those of an earlier study showing that individuals who *were* found to be at increased genetic risk of melanoma became much more diligent about protecting themselves from sun exposure, conducting self-screening exams, and seeing their dermatologist regularly. The logical conclusion is that genetic screening coupled with proper counseling produces a beneficial response for the person who is screened for risk of melanoma, no matter what the results.

Counseling can be provided by a doctor, nurse practitioner, physician's assistant, genetic counselor, psychologist, or other healthcare professional and usually takes place at a doctor's office or a clinic. A few direct-to-consumer companies are now also offering some degree of counseling via email or by telephone.

The decision about how and where you'd like to receive your counseling should be based on your own personal preference and what you believe would make you feel most comfortable. But, however you choose to receive it, counseling by a healthcare professional at the time of genetic screening reduces fear; dispels misconceptions; and increases understanding of the test, the analysis, and the results.

CALCULATION OF GENETIC RISK

Most personal genomics companies use similar formulas for calculating a person's genetic risk of disease.

For rare diseases, the risk assessment is straightforward. Basically, there are three possibilities: you are a carrier, you most likely have the disease, or you are not a carrier and not affected. For common diseases, risk is assessed by combining the risk associated with all the variants you have for a disease and possibly also integrating nongenetic factors—such as whether you are a smoker—into the formula. Often, the risk for common diseases is most easily conveyed to you as a single lifetime risk value.

One criticism of calculating the risk for common diseases is that some of the variants increase risk only marginally. For example, a variant that increases the risk of a disease by 50 percent may sound quite serious, but if the lifetime risk of the disease for everyone in the general population is 2 percent, that means the lifetime risk for people who have this variant is only 3 percent. When viewed in that context, a 50 percent increase in risk no longer seems so ominous.

For this reason, it is important for you to know both the lifetime risk value for the general population and your own lifetime risk value based on your genetic makeup. Comparing the two is what allows you and your physician to determine whether your risk is substantial and requires action.

Although some geneticists dismiss variants that increase risk of a disease by only a small amount, I still find them to be of tremendous value primarily because they usually don't occur in isolation. What this means is that a person may have a few different variants associated with the same disease, each of which increases risk only marginally but that, in combination, may cause a clinically significant increase.

ARE PRELIMINARY STUDIES USEFUL?

The validation of any scientific inquiry, including the genetic research used in personal genomics and predictive medicine, depends on the ability of independent researchers to replicate the findings. Therefore, at least two or more studies showing similar results are usually needed before we can say whether a specific genetic variant is truly associated with a disease or trait. If only a single study has shown the association, the results are considered *preliminary* and are said to *require replication*.

As both a medical doctor and a genetics researcher, I believe that some preliminary results—if the research study has been rigorous and held to the highest scientific standards—may hold value. In some instances, for example, the variant may be associated with a rare disease or trait that has been studied by only a single research group. In such cases, the preliminary data these studies provide may be the only information available. In other instances, researchers may have conducted such a thorough examination of the association between a specific variant and a disease that the preliminary results can be considered reliable. As long as the experimental nature of preliminary associations are disclosed and understood by both you and your doctor, I don't believe that such findings should automatically be dismissed or ignored.

Whether you and your doctor choose to act on preliminary results or to use only fully replicated data is, of course, a personal decision and may depend on what kind of risk you are assessing. If you are interested in your risk of a heart attack, for instance, you may choose to use only well-established data. But if you have a rare type of leukemia and your physician is deciding between two possible medications, there may be only one study comparing those medications, and if that study indicates that, based on your genetic makeup, one medication will be much more effective than the other, the two of you may decide to act on this information even though the evidence is still preliminary.

Generally, the decision of whether to act on preliminary data is likely to depend on the seriousness of the preventive measures associated with the disease. If, for example, a preliminary result showed that eating more vegetables would lower your risk of a heart attack, taking that action would be both benign and reversible; thus the potential benefits would far outweigh the negligible risk. If, on the other hand, the suggested preventive measure was a major surgical procedure, then you and your doctor may be more cautious about proceeding on the basis of preliminary evidence alone.

Generally, the decision of whether to act on preliminary data is likely to depend on the seriousness of the preventive measures associated with the disease.

The point is that it is extremely important that the personal genomics company you choose be very clear about whether the results it reports are based on preliminary or replicated research, so you and your physician can make an informed decision about whether to act on them.

HOW VALID ARE THE RESULTS?

There has been considerable controversy about the clinical usefulness of genetic testing and analysis. For example, two studies published in 2008 in the *New England Journal of Medicine* have been cited by the media as showing that genetic screening, as one headline put it, "Offers Little Help for Diabetes Prediction." Unfortunately, this statement is grossly inaccurate. First, both studies actually showed that genetic screening *does* lead to a better prediction of a person's risk of diabetes than other current methodology (such as family history, weight, or body mass index). Most likely, the critics simply misinterpreted the findings, which stated that genetic screening "slightly" improved risk prediction by latching onto the word *slightly* and interpreting it to mean that the overall usefulness of such screening is questionable.

In addition, they missed the fact that, while all the physicians in the study made sure to take a complete family history and calculate the body mass index for each and every patient because that was part of the study protocol, community-based physicians might not actually do this as thoroughly or as often. Many community-based doctors are so overloaded with work that they may forgo a thorough and time-consuming assessment of diabetes risk and focus their limited time on treating more acute and immediate problems such as an infection or a painful joint. And even if the doctor did inform the patient that his or her body mass index indicated an increased risk for diabetes, the patient might ignore the warning because it was nonspecific, thinking "Yeah, I know you say that to everyone who's on the larger side." Genetic information, on the other hand, is so personalized that the patient is more likely to take it seriously and comply with the doctor's preventive recommendations.

And, finally, a very important concept the critics overlooked was that genetic screening can predict diabetes risk decades *earlier* than any other method. An obese adult may have been quite fit as a child, and only genetic screening would have alerted the child and his or her parents to institute nutrition and fitness goals that could prevent or reduce the risk for diabetes later in life.

The point isn't that physicians should start to use genetic screening *in place of* other ways to predict diabetes risk but rather that genetic screening is an efficient way to offer more personalized and focused care. Not only does genetic screening allow the physician to predict disease risk sooner and more accurately but it also can provide genetically tailored information about how to prevent and treat that disease.

Genetic screening can provide recommendations about prevention and treatment that are specifically tailored to the person's genetic makeup.

One criticism that is more scientifically astute has been levied by David Goldstein, a Duke University researcher who is well respected in the genetics field. In an article published in 2009, Goldstein points out that some diseases and traits may be dictated by hundreds or even thousands of rare genetic variants and that most personal genomics companies today test for only a few of the most common variants and thus do not uncover the full picture. My answer to this criticism is twofold: First, there are a few personal genomics companies that do, in fact, test for almost all *known* variants associated with a disease regardless of how common or rare, and these companies are, therefore, able to integrate a significant amount of information into their risk calculations. And, second, while its true that there could be many variants associated with a disease that we have yet to discover, I believe the amount of knowledge we have right now is *more than enough* to provide both patients and physicians with actionable information that is capable of preventing disease.

Over time, the number of variants known to be associated with a specific disease or trait will certainly grow, but that doesn't mean we can't or shouldn't use the information we currently have available to us. Although some critics look at the data and see what is missing, I prefer to look at the data and see the tremendous amount we already have.

While we certainly don't yet have all the answers, it is important to remember that medical science is constantly in flux and always evolving, advancing, and improving. Medical history has shown that what we know right now is different from what we will understand tomorrow—but that does *not* negate the usefulness of the knowledge we have today. We do not need to have every single piece of the puzzle in place in order to see the picture, because in medicine, no puzzle is *ever* complete. Therefore, we must go forward today and use genetic information to the best of our abilities so that we can protect and preserve our future health.

WHAT YOU SHOULD KNOW ABOUT REGULATION

Globally, regulation of the personal genomics industry is quite lax, and in the United States, the Food and Drug Administration (FDA) does not currently regulate it. The only oversight of personal genomics we have at present comes from the Centers for Medicare and Medicaid Services (CMS), the federal agency tasked with enforcing clinical laboratory standards known as the Clinical Laboratory Improvement Amendments (CLIA).

Any lab in the United States that performs clinical testing is required to receive CLIA certification. The key word here is *clinical*, because a facility that does not perform tests for human medical use, such as a research lab or a veterinary lab, does not require CLIA certification. And because some personal genomics companies state that their services are not meant to be used for medical purposes, some of them use labs that are not CLIA certified.

CLIA certification sets minimum standards to ensure that the facility is run by well-trained, capable people and that the tests they run are performed to a certain, replicable proficiency standard. It does not, however, certify the validity of the tests nor does it provide oversight of the genetic analysis that occurs after laboratory testing.

Even though laboratories outside of the United States are eligible to receive CLIA certification, many choose a more global regulatory oversight from the International Organization for Standardization (ISO). Like CLIA certification, ISO accreditation (specifically, ISO-15189) requires a set of minimum laboratory standards and protocols but does not regulate or certify the validity of the tests themselves or of the analysis.

Given the fact that laboratory oversight is extremely important for ensuring reputable testing techniques, I highly recommend purchasing anything related to genetic testing only from a company that clearly states (and is able to provide supporting documentation) that the laboratory it uses for *all* its testing is either CLIA certified or ISO-15189 accredited.

During the time I was writing this book, I researched a Canadian company that offers very low-cost genetic screening and uses a lab based in the U.S. Southeast. I wasn't able to find any CLIA information on the company's website, so I emailed them and received a very friendly reply stating that yes, its lab was CLIA certified. However, I was familiar with the laboratory in question, and I knew that it was *not* certified at that time. So I emailed the company again and asked them to provide the lab's CLIA certification number (which should always be available to anyone who asks). I received an email response saying that someone would get right back to me with it, and I never heard from the company again.

Laboratories that are approved by CLIA or ISO usually charge more for their services than those that are not, so personal genomic companies that use these higher-cost, higher-quality labs are almost always very forthright about providing certification information to potential customers and in their marketing materials.

Although the federal government may not regulate the specific tests and analyses offered by personal genomics companies, 25 state agencies do have specific restrictions. For example, both California and New York require any company offering genetic screening to apply for special licensure from the state's Department of Health. To be licensed, the company must submit documentation supporting the scientific validity and accuracy of the testing it plans to offer. Combined with CLIA certification, state licensing helps ensure that your DNA sample will be processed professionally and correctly at the lab and that the association between a specific disease and the genetic variants being analyzed is supported by the scientific literature.

GENETIC INFORMATION AND CONFIDENTIALITY

An understandable concern many people have is that their genetic information might not remain confidential and could potentially be misused. One such fear concerns genetic discrimination—the possibility of being treated differently, such as by an insurance company or an employer, based on personal genetic information.

In the past, many people have chosen to forgo genetic screening rather than run the risk of such discrimination. However, in 2008, Congress passed the Genetic Information Non-discriminatory Act (GINA) to prevent both health insurance companies and employers from discriminating based on genetic information. It is now *illegal* for a health insurance company to increase your premiums or refuse you coverage based on genetic information. Nor are employers allowed to base hiring, firing, or promotion decisions on an employee's (or prospective employee's) genetic makeup.

GINA prevents health insurance companies and employers from discriminating based on a person's genetic information.

Still, I believe it is extremely important that you keep your genetic information confidential. First, GINA has yet to be tested in court and, therefore, should not be fully relied on. Second, GINA contains some significant loopholes, the most glaring of which is that while it protects you against discrimination by health insurance companies, it does *not* protect you against discrimination by life insurance, disability insurance, or long-term-care insurance providers. Nor does it protect against discrimination by employers if the company has fewer than 15 employees.

The primary measure you can take to ensure the confidentiality of your genetic information is to verify that the personal genomics company you are thinking of using is reputable and offers an assurance of confidentiality. Neither your name nor any other identifying information should ever be linked to your genetic information, either in the personal genomics company's database or in your genetic report. You should be assigned a means of identification other than your name, such as a confidential identification number (CIN), which is then used in place of your name. This way, even if a third party somehow managed to access your genetic information, there would be no way to associate the information with you.

Always verify that the personal genomics company you are thinking of using offers an assurance of confidentiality.

It is also important to realize that some personal genomics companies state that they may use your genetic information for research purposes. This terminology is quite broad, however, and may mean that the company can either license access to or sell your genetic information to a third party, such as a research group or another company. Some personal genomics companies may even *require* that you allow your genetic information to be used for these purposes; others make the choice fully optional. At the very least, you should always make certain the company states that no personally identifying information will be included if your genetic information is used.

Concern for your confidentiality should not stop you from pursuing genetic screening, but it should be one of the factors you consider when deciding which company to use.

HMOS, HEALTH INSURANCE COVERAGE, AND GENETICALLY TAILORED INSURANCE

While HMOs and health insurance companies do at times reimburse for genetic testing of rare diseases if there is a clear indication that it may run in a family, they do not normally cover genetic screening for common diseases. The most likely reasons for this are, first, that the tests are new and, therefore, the insurers have yet to evaluate them fully and, second, that some common diseases take decades to manifest while many HMOs and health insurance companies are only interested in preserving your health for the next 10 years. Most people change their coverage frequently throughout their life (when, for example, they change jobs), so the insurer you have today is most likely different from the one you'll have in 10 years. As a result, the insurers don't think it's cost effective for them to cover the prevention of a potential illness that most likely won't manifest until after you've left their plan.

This, of course, highlights the fact that neither HMOs nor health insurance companies have much incentive to truly support prevention, even though they like using the words "preventive care" in their advertising campaigns. Because of this lack of coverage, the majority of preventive services, including genetic screening, remain primarily out-of-pocket expenses even though their cost is almost always substantially less than the cost of treating a disease once it manifests. As Benjamin Franklin famously stated, "An ounce of prevention is worth a pound of cure."

Genetically tailored insurance coverage may be an option in the near future.

In the near future, however, specific insurance options may be offered to people who, through genetic screening, are found to be at increased risk of a disease. As we discussed in the last section, a federal law called GINA protects against health insurance companies discriminating (such as deciding whether to accept someone into their plan or what premiums that person will have to pay) based on genetic information, but it doesn't prohibit companies from using genetic screening to identify either new or existing clients who may benefit from specialized insurance products. Therefore, these companies may soon start to offer insurance tailored to a person's genetic makeup that would cover both preventive measures and treatment should diseases they are predisposed to ever manifest.

READ THE FINE PRINT

As with anything you purchase, you need to know what you're getting, and sometimes personal genomics companies use language that's designed to mislead.

One misrepresentation I've encountered quite often is that a company may advertise the number of genetic variants it *tests* for, but that doesn't mean it uses them all in its *analysis*. For example, the company's website or marketing materials might state, "We read more than 500,000 locations in your genome," "We capture over 900,000 markers in the genetic code," or "We test over 1 million places in your DNA," but notice that the wording here (*read*, *capture*, and *test*) does not ever mention *analyze* in relation to disease risk.

As with anything you purchase, you need to know what you're getting.

Most companies actually hide the number of genetic variants used in their *analysis* of each disease because the number may be quite small. A company that tests for more than 1 million places (that is, genetic variants) in your DNA may actually use only 100 or 200 of those in their analysis. And because they may test for 40 or more diseases and traits, their assessment of your risk for each disease is determined by only 1 to 5 variants. This means

It's important to determine how many variants are actually used in the analysis of disease risk, not just the number they test for.

they are using as little as 0.01 percent of the genetic information they obtain from their testing in their analysis. The few variants that they do analyze are usually either the ones that occur with the highest frequency and are thus the most likely to be detected or the ones that just happened to be on the specific array they are using to conduct their testing. Unfortunately, however, this may mean that the company is not analyzing dozens or even hundreds of other variants also associated with the disease.

Alzheimer's, for example, has two forms: sporadic, which is the common disease most people are familiar with, and familial, which is a rare disease. Many personal genomics companies calculate the risk of sporadic Alzheimer's disease from a *single* genetic variant and don't provide any testing or analysis for the familial form. There are, however, many other genetic variants known to be associated with sporadic Alzheimer's disease and more than a hundred associated with the familial form that will therefore be overlooked.

The bottom line is that not all companies are able to test for and/ or analyze a large number of variants that are known to be associated with a disease. Even a personal genomics company that offers full genome sequencing (and therefore tests for all variants) may only use a very small amount of that information in its analysis of disease risk.

Unfortunately, this kind of important information often isn't provided unless you specifically ask for it.

THE FIVE MOST IMPORTANT QUESTIONS TO ASK— AND THE ACCEPTABLE ANSWERS

You should be able to find out a lot of useful information just by reading a company's website. For example, most companies that use CLIA-certified labs usually make sure to state this fact clearly. The website should also include a telephone number that you can call for further information.

So, whether you get the information from a website or from a customer service representative by email or on the phone, here are the questions you should ask and the answers you want to hear.

1. **Laboratory:** Is the laboratory you use for *all* lab work currently CLIA certified or ISO-15189 accredited?
   - **Acceptable answer:** Yes, the company should also be able to provide its CLIA certification number or ISO-15189 accreditation information so you can verify that certification.
   - **Unacceptable answer:** No, "It's currently in process," "We don't need that for what you want," or "We are certified but can't provide documentation in writing."

2. **Types of diseases:** Do you conduct genetic testing and analysis for rare diseases (referred to as monogenic diseases), for common diseases (referred to as multifactorial diseases), or for both? How many diseases in each of those categories do you test for, and do you offer panels?
   - **Acceptable answer:** It depends on what you want to be screened for, but if you are looking for both risk and carrier information for as many diseases as possible, the answer should be both. And if the company offers testing for 20 or more diseases and traits, it should offer at least a few different panels (otherwise the volume of information you receive may be too overwhelming and thus of little value).
   - **Unacceptable answer:** No reply or an evasive reply such as, "We test for many important diseases." If the company spokesperson isn't forthright with you, you might want to consider looking for another company.

3. **Comprehensiveness of analysis:** How many genetic variants do you *analyze* (not just test for, but analyze) when you calculate disease risk?
   - a. How many do you analyze for Alzheimer's disease?
   - b. How many do you analyze for cystic fibrosis?
   - c. How many do you analyze for Tay-Sachs disease?
   - **Acceptable answers:** The number of genetic variants analyzed is key, *not* the number of genes the company tests for. All companies may test the same gene for cystic fibrosis, but some companies may analyze only

a single variant within that gene, whereas others may analyze hundreds. Common diseases are usually associated with 10 or more genetic variants. For rare diseases, there are usually 25 or more, sometimes even hundreds, that are known to cause the disease—possessing just one of those variants can cause a person to either be a carrier or have that rare disease. Therefore, if a company provides risk information about 50 diseases, the *analysis* could include 500 different genetic variants. And if risk information is provided for 200 diseases, the *analysis* will likely involve more than 2,000 variants.

    a. *For Alzheimer's disease:* The sporadic form alone is associated with at least 5 genetic variants; testing for both the sporadic and the familial forms involves more than 100 genetic variants.

    b. *For cystic fibrosis:* There are more than 10 likely variants; a thorough evaluation would involve more than 1,000.

    c. *For Tay-Sachs disease:* There are at least 6 likely variants; a thorough evaluation would include more than 100.

- **Unacceptable answer:** Giving you the total number of variants the company *tests for* rather than the number it uses in its analysis of disease risk. And while a company may say it uses 50,000 genetic variants in its analytical process, 99.9 percent of those variants may relate only to genealogical testing and not to disease testing. Therefore, it's important that the company tells you the number of genetic variants it analyzes when calculating *disease risk.*

4. **Actionability of information provided:** Will the results be presented to me along with genetically tailored prevention and treatment information, and will I be able to understand the genetic report even if I don't have any knowledge of genetics?

- **Acceptable answer:** Whether you want to receive genetically tailored prevention and treatment information along with your risk of a disease is a personal choice. And the type of report that will be most useful to you fully depends on your level of comfort in understanding genetics and medicine. If you have a medical background or feel confident that you can appropriately research diseases and prevention on your own, actionability and the clarity of the genetic report may not important to you. But if you would like to ensure that the report is both actionable and straightforward, the acceptable answers are yes and yes.

- **Unacceptable answer:** "We will provide you with links to other websites where you can read more about the disease" or "You may need to take our tutorial on genetics before you're able to fully understand our genetic report and your results."

5. **Format of results:** Will the results be delivered to me via email or on a website, or will I receive printed reports for both me and my physician? Do you provide a way for me to consult with a doctor or other healthcare professional who can help me understand my results?

- **Acceptable answer:** Again, this is a matter of personal preference. Some people are comfortable receiving information via email or on a website, whereas others prefer to have a printed report. Having a physician involved in the process is also a personal choice, although physicians usually prefer to have a printed report. Remember too that your physician may prefer to use a company he or she is familiar with, so if you want your physician to be involved in the process (reviewing the results and possibly acting on that information), it might be best to check with him or her first.

- **Unacceptable answer:** The results are delivered to you on a website, but you will not be able to download or print them. (No company should be able to dictate or limit what you can and cannot do with your own genetic information.)

LIST AND COMPARISON OF COMPANIES PROVIDING GENETIC SCREENING

The genetic screening services provided by the companies listed in the table below may differ significantly. Some services are à la carte (each disease is sold separately), some are panel based, and others are all in one (all diseases are tested for at once).

NAME	WEBSITE	OFFERED THROUGH	TESTING SOLD AS	ACTIONABLE GENETIC REPORTS	COMMON DISEASES	RARE DISEASES	NUMBER OF VARIANTS ANALYZED PER DISEASE	CLIA CERTIFIED	HEALTHCARE PROVIDER CONTACT
Ambry Genetics	ambrygenetics.com	Healthcare professionals	À la carte	No	No	Yes	Many	Yes	Yes; MD, in-person
Existence Genetics	existencegenetics.com	Healthcare professionals	Panels	Yes	Yes	Yes	Many	Yes	Yes: MD, in-person
GeneDx	genedx.com	Healthcare professionals	À la carte	No	No	Yes	Many	Yes	Yes: MD, in-person
Genzyme Genetics	genzymegenetics.com	Healthcare professionals	À la carte	No	No	Yes	Many	Yes	Yes: MD, in-person
Personalized Medicine Group	hhdocs.com	Company offices	À la carte	Yes	Yes	Yes	Many	Yes	Yes: MD, in-person
23andMe	23andme.com	Direct via Internet	All in one	No	Yes	No	Few	Yes	No

ABBREVIATIONS: MD = medical doctor; GC = genetic counselor.
Data collected April 2009.

LIST AND COMPARISON OF COMPANIES PROVIDING GENETIC SCREENING

NAME	WEBSITE	OFFERED THROUGH	TESTING SOLD AS	ACTION-ABLE GENETIC REPORTS	COMMON DISEASES	RARE DISEASES	NUMBER OF VARIANTS ANALYZED PER DISEASE	CLIA CERTIFIED	HEALTHCARE PROVIDER CONTACT
deCODEme	decodeme.com	Direct via Internet	Panels	No	Yes	No	Few	Yes	Yes: GC, phone/email
DNAdirect	dnadirect.com	Direct via Internet	À la carte	Yes	Yes	No	Few to many	Yes	Yes: GC, phone/email (for a fee)
Navigenics	navigenics.com	Healthcare Professionals	All in one	Yes	Yes	No	Few	Yes	Yes: MD, in-person
Pathway Genomics	pathway.com	Healthcare Professionals	All in one	No	Yes	No	Few	Yes	Yes: MD, in-person

ABBREVIATIONS: MD = medical doctor; GC = genetic counselor.
Data collected April 2009.

If a company provides genetic testing through a healthcare professional, its services implicitly involve face-to-face contact with a physician and may also involve consultation or counseling with other healthcare providers, such as a genetic counselor, a nurse practitioner, or a physician's assistant who might be working with the physician.

It is also important to point out that this list is not all-inclusive and the companies listed in the table are primarily those that provide the option of screening for many different diseases and disease predispositions. There are other companies that offer screening for just one or a few diseases. I have not listed these companies because of the limited scope of genetic information they supply. There are also hundreds of other companies that provide genealogical and paternity testing.

Now that you understand the important concepts and options surrounding genetic screening and predictive medicine, it's time to put all this information into action. The chapters in Part II discuss exactly how predictive medicine can help you outsmart your genes and prevent some of the most widespread and dangerous diseases currently plaguing our society.

Outsmart Your Genetic Destiny with Predictive Medicine

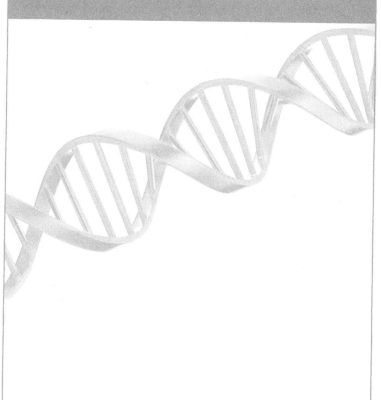

5

Do These Genes Make Me Look Fat?

MISCONCEPTION: If you eat right and exercise, you'll be able to lose weight and stay slim.

FACT: For many people, eating healthy and exercising will *not* work, and often this is because of their genes. Your genes not only dictate a significant portion of your metabolism and weight but also control whether specific diets and forms of exercise will be effective in reducing and controlling your weight.

You've heard it before: Obesity has reached epidemic proportions throughout the developed world. You've seen the never-ending supply of books, products, and contraptions designed to promote weight reduction, and you've also heard disappointed consumers lamenting that a particular diet or exercise routine "just didn't work for me." The key here is the *for me* part, because it implies that the same plan actually has worked *for someone else*. The fact that a specific diet or exercise didn't work for you may very well be due to the fact that

A specific diet or exercise may not work for you, even though it works for other people, because of your unique genetic makeup.

those interventions were not personally tailored to your genetic makeup. This is extremely important information because our genes are responsible for determining approximately 70 percent of our weight, with nongenetic factors, such as lifestyle choices, determining the remaining 30 percent.

The good news is that because we can now examine your entire genetic makeup, we can formulate a genetically tailored fitness and nutrition plan that is much more likely to work for you. Using this genetic insight, you'll be able to focus on the diet and exercise regimen that will most likely benefit you while avoiding those that are less likely to help. A preliminary study comparing people who used a

Genetic screening can now provide you with a genetically tailored fitness and nutrition plan.

genetically tailored nutrition plan to those who didn't found that genetic tailoring helped people achieve more significant weight reduction over longer periods of time.

In this chapter first I'll discuss obesity and how specific genetic variants are partly to blame for why many people become overweight. Then I talk about ways in which diet and nutrition can be tailored to your specific genetic makeup in order to facilitate weight loss. At the end of the chapter, I cover the ways in which fitness and athletics can be personalized to your specific genetic makeup and how we can now start to use genetic screening to assess whether weight-loss medications and even weight-loss surgery will be effective.

THE FAR-REACHING CONSEQUENCES OF OBESITY

In 2009, the Centers for Disease Control and Prevention (CDC) issued a report on the obesity epidemic, stating that more than *one third of all adults* in the United States, representing more than 72 million people, are obese. The report also stated that "since 1980, obesity rates for adults have doubled and rates for children have tripled. Obesity rates among all groups in society—irrespective of age, sex, race, ethnicity, socioeconomic status, education level, or geographic region—have increased markedly." In the same year, another report written by the chair of Emory University's School of Public Health found that obesity is responsible for at least 20 percent of the rise in healthcare spending over the last two decades and that "if the prevalence of obesity were the same today as in 1987, healthcare spending in the United States would be about $200 billion less each year." From a global perspective, the World Health Organization has stated that more than 2 billion people worldwide are currently obese and that by 2020 *more than 60 percent of all global diseases* will be chronic diseases associated with obesity.

Based on these statistics it is clear that combating obesity and improving each person's nutrition and fitness have positive consequences that go far beyond aesthetics. Obesity, suboptimal nutrition, and a sedentary lifestyle all lead to significantly increased risk of cancer, Alzheimer's, heart disease, high blood pressure, strokes, fertility problems, diabetes, depression, sleep abnormalities, difficulty healing from injuries, back problems, and arthritis. Being overweight or obese over many years has also been shown to cause premature aging and outright atrophy of the brain. All these problems lead to decreased work productivity, increased rates of chronic and acute diseases, and an unsustainable strain on our healthcare system—not to mention the psychological turmoil and reduced self-esteem inflicted on those who have been struggling for decades to control their weight.

NEW WEAPONS IN THE BATTLE OF THE BULGE

New weapons—such as more effective diets, exercise programs, medications, and even surgical procedures—are needed to augment the armament we already have to combat obesity. Predictive medicine and genetic screening have an extremely important role to play in the use of all these tools.

Genetic screening will not make you thin and fit overnight. And while it can provide some truly novel insights that can help you arrive at your own personalized nutrition and fitness regimen, you're still going to need the motivation and fortitude to follow through and put that information into action. That said, however, if you are determined to give it your best shot, genetic screening can help you achieve your short- and long-term goals. Just as you can outsmart the genes that predispose you to disease, you can also use genetic screening to outsmart the genes that predispose you to obesity.

Genetic screening can empower you with the insight needed to outsmart the genes that predispose you to obesity.

It's been suggested that if people find out they have a genetic predisposition to obesity they will be less motivated to adhere to a weight-loss program. However, a recent study conducted by researchers at the Johns Hopkins University and the University of Vermont found that this concern was actually unwarranted. When people found out that they were genetically predisposed to obesity, their confidence in their ability to lose weight did not decrease. In fact, they gained *more* confidence in their ability to control their weight, which is known as *weight management self-efficacy*. The researchers noted that increased self-efficacy has already been linked to increased adherence to weight-loss and weight-management treatments. Therefore, genetic testing for a predisposition to obesity might actually increase people's adherence to a weight-loss program.

To date, the treatment of obesity has not accounted for the genetic variations that make each of us unique. Now, however, scientists and physicians are at last beginning to understand the genetics of weight regulation and what can be done to solve the obesity problem. With that understanding, they can tailor diets, exercise programs, medications, and other treatments to each person's genetic makeup. It's time to make the fight against obesity personal.

THE WEIGHT OF YOUR DNA

Weight gain usually occurs for a single reason: You are consuming more calories than you are expending. This may happen because you are eating too many calories, because you aren't burning enough calories, or a combination of the two.

Our genes control our weight by affecting this in and out balance. For example, some genetic variants cause people to be hungry more often, and since they eat more, they gain more weight. Other genetic variants cause people to have a slow metabolism, which means they burn calories at a slower rate. And there are even variants that are known to cause fat to build up faster in certain areas, such as in our thighs or in our sides, creating those infamous love handles.

In terms of exercise, there are variants that cause you to burn fewer calories than others even though you perform the same amount of physical exercise. And even more genes control your muscles' ability to tolerate exercise so that variants in these genes may result in your leading a more sedentary lifestyle.

OF OBESE MICE AND MEN

THE STORY of how the first obesity-related gene was identified begins all the way back in 1950 when researchers at the Jackson Laboratory in Bar Harbor, Maine, found that one of their strains of laboratory mice was extremely fat and so hungry that they ate virtually all the time. No matter how much they ate, they still wanted more food.

It was not until 1994, however, that a research team at the Rockefeller University in New York City discovered that the reason these mice ate so much and were so large was because of variants in the gene responsible for producing the protein leptin.

continued

Leptin is produced by fat cells and is secreted into the blood. It then circulates throughout the body and signals the brain that it's time to turn off the feeling of hunger. Therefore, the leptin gene is responsible for producing our body's own appetite suppressant.

Variants in this gene, however, can cause it to malfunction. As a result, the body can't produce adequate amounts of leptin, the brain never gets the off signal, and the mouse's appetite becomes constantly and overwhelmingly voracious. When the obese mice were injected with the correct amount of leptin, they stopped overeating and lost weight! The scientists, understandably, believed they'd found the cure for obesity.

Unfortunately, however, when they began to study the leptin gene in obese humans, they found that variants within this gene were not a common cause of obesity. And when one drug company administered leptin to obese people, only a small percentage lost significant weight.

Since the discovery of the leptin gene, however, researchers have made great strides in discovering other genes that are much more common causes of obesity in humans.

SIZING UP A FAT GENE

Yes, I know a guy is never supposed to utter these words, but I'm going to say them in the name of science: "These genes really *do* make you look fat."

In 2007, the first commonly occurring fat gene was identified. Variations in the fat mass and obesity-associated (*FTO*) gene were found to correlate with increased weight in both children and adults. These variants are found in 10 to 50 percent of people in most ethnicities, including Caucasians, Asians, Hispanic Americans, and African Americans. If we consider just one of these possible variants, children and adults with one copy of this variant have about a 30 percent increased risk of developing obesity, and those with two copies have close to a 70 percent increased risk. In the population as a whole, variations in the *FTO* gene are thought to be responsible for a significant proportion of obesity.

In its normal state, the *FTO* gene carefully regulates the amount of fat our bodies store by controlling appetite. Variants in this gene cause people to feel hungry even if they've eaten enough food, leading to increased consumption. A 2008 study in the *New England Journal of Medicine* found that individuals with these obesity-causing variants not only eat more but also have a tendency to eat foods that contain more calories. Variants in this gene, therefore, pack a true double whammy.

Variants in the *FTO* gene contribute to obesity not only by causing people to eat more but also by causing them to eat foods that contain more calories.

While increasing physical activity (such as by starting an exercise routine) will result in significant weight loss in individuals with these variants, this lifestyle modification does not treat the true cause of the weight gain. To outsmart genetic variants within the *FTO* gene, people need to concentrate on their diet and controlling hunger. There are numerous ways to do this, including behavioral modification programs run by a psychologist or psychiatrist, weight-loss programs that focus on balanced eating, and appetite-suppressant medications that may help counteract hunger.

OTHER FAT GENES

Each year researchers are discovering more genetic variants associated with weight gain and obesity. For example, in 2008, variants in the *TUB* gene have been found to trigger weight gain in women later in life. This gene is active in areas of the brain that control food consumption, and specific variants have been found to cause postmenopausal women to eat more simple carbohydrates, which means that their diet will provide a higher glycemic load. This dietary change is thought to be one reason for the increased weight observed in women with *TUB* variants.

ARE YOU OVERWEIGHT OR OBESE?

THE BODY mass index (BMI), a ratio based on the relationship between height and weight, is commonly used to figure out if a person is underweight, normal, overweight, or obese. If you want to find out your BMI, there are easy-to-use calculators on a number of websites. Simply plug in your weight and height, and you'll be given your BMI. An easy-to-use BMI calculator, along with an explanation of the results, can be found at www.Outsmart YourGenes.com/BMI.

While there are many other fat genes, the final one we'll talk about here answers the question: With so many obesity-related genetic variants, what happens if I have more than one?

The *MC4R* gene contains variants that have been associated with increased weight, obesity, and a predisposition to having a larger than average waistline. Scientists believe that the gene is responsible for modulating eating behavior and that when variants cause it to malfunction, the result is an overwhelming need to eat. Children and adults with *MC4R* variants tend to snack more often, eat foods with higher fat content, and also tend to eat much larger quantities of food.

Overall, variants within the *MC4R* gene are thought to be responsible for causing obesity in about 5 percent of obese individuals. While variants in this gene are definitely not as common as those in the *FTO* gene, they are still very important in causing obesity when they do occur.

Women with variants in their *MC4R* gene have, on average, a BMI that is close to 10 points higher than women without those variants, while men have a BMI that is 4 points higher. It is thought that one of the male hormones may blunt some of the effects, thus accounting for the sex differences in BMI.

In 2009, researchers from France published a study that specifically investigated the *MC4R* and *FTO* genes. They found that having variants in both genes increased the risk of obesity considerably more than either variant on its own.

SKINNY GENES

WHAT ARE some of the physical features that come to mind when someone mentions a Scandinavian woman? Usually, it is blond, tall, and . . . thin. Scandinavians are known for their tall, thin physique, and although a healthy lifestyle may play a role, there are also genetic forces at work.

Some Scandinavians are genetically predisposed to being thin. Variants in the *GPR74* gene, carried by 1 in 20 Scandinavians, have been found to significantly increase the body's ability to break down fat. Because of this, people with these variants are predisposed to having lower body fat and slimmer waistlines.

In addition to the *GPR74* variant, there are other so-called skinny genes, many of which are involved with either fat storage or rate of metabolism.

PERSONALIZED PREVENTION AND TREATMENT

If you discover that you are genetically predisposed to being overweight or obese what can you do about it? The answer is a lot.

Aside from anything else, finding that you have such a predisposition can relieve you of the self-blame and guilt you may have been harboring, which may, in turn, motivate you to turn your attention to outsmarting the true culprit. And your genetic makeup also contains the answers to how you can best lose weight and keep it off.

GENETICALLY TAILORED NUTRITION

The emerging field of nutrigenomics (the genetics of nutrition) examines the ways in which your genes interact with what you eat and drink. With such knowledge, healthcare professionals, nutritionists, and fitness trainers can start tailoring your nutrition and diet precisely to your genetic makeup. A very interesting study published in 2007 found that the "addition of nutrigenetically tailored diets resulted in better compliance, longer-term BMI reduction and improvements in blood glucose levels." Everything from the most effective weight-loss diet to what vitamins and foods your body needs most can now be determined through genetic screening.

Nutrigenomics allows you to tailor your nutrition and diet precisely to your genetic makeup.

Researchers from the University of Connecticut have published preliminary studies about specific genetic variants that determine whether a low-fat diet or low-carbohydrate diet will be more effective in allowing a person to lose weight. They found three genes (*PFKL*, *HNMT*, and *RARB*) have variants that increase the likelihood of losing weight with a low-fat diet and five genes (*LIPF*, *CETP*, *AGTR2*, *GYS2*, and *GAL*) have variants that increase the likelihood of weight loss with a low-carb diet.

But, you may ask, if my weight is determined by the number of calories I take in versus the number I expend, why should it matter where those calories come from? To answer that question, let's look at the *LIPF* gene, which is responsible for breaking down the fat you consume once it enters your stomach. A genetic variant present in about 30 percent of Asians, 15 percent of Caucasians, but much less often in people of African descent, causes the gene to malfunction so that fat is not broken down very well in the stomach and, therefore, is not as well absorbed by the body.

Eating a low-carb diet usually means that a person will be increasing the amount of fat they consume because instead of eating foods high in carbs, they switch to foods that are higher in fat, such as meat. Therefore, if a person has this variant in their *LIPF* gene, a low-carb diet is more likely to work because the person would be actively cutting down their calories from carbs while their body would naturally be cutting the amount of fat it absorbed (even though the person would be eating more fatty foods, the body wouldn't be absorbing all of it). These two actions work together and are more likely to lead to successful weight loss. Conversely, if a person doesn't have this variant, their body will be absorbing a lot more of the extra fat they'll be consuming on a low-carb diet, and, therefore, they will be less likely to experience the same degree of weight loss.

Variants in the *PFKL* gene, which is involved in converting carbohydrates into energy, can also affect your chances of losing weight on a low-fat, high-carbohydrate diet. One variant, present in approximately 1 out of 7 people, causes almost no loss of body fat on this type of diet, but another variant, the more prevalent one, causes a substantial loss of body fat and weight on the same diet. Even though the two variants occur in the same gene, they have different effects on the body.

EAST MEETS WEST: GENES AND NUTRITION

PEOPLE OF all ethnicities and cultures are now living in places other than their country of origin. For some people, especially Asians who are now living a more Westernized lifestyle, this has resulted in an astounding increase in obesity. Because of this, researchers have started to investigate whether genetic differences and a Westernized, high-calorie diet are influencing this observed weight gain.

It's been known for a while that Japanese Americans who adopt a Westernized diet go on to develop obesity and type 2 diabetes at a much higher rate than their relatives who still live in Japan and eat a traditional Japanese diet. In 2008, researchers at Hiroshima University in Japan published a preliminary

continued

study showing that a variant in the *ITGB2* gene, which is present in almost 30 percent of Japanese people, is associated with an increased risk of obesity only among those who live in the United States and follow a Westernized diet. Japanese with this same variant who follow a more traditional diet do not have an increased risk of obesity. The authors conclude that the *ITGB2* variant may be one reason why people of Japanese descent become obese when exposed to a Westernized diet.

The *ITGB2* gene plays a role in dictating a person's metabolic rate, and it appears that, over thousands of years, certain genetic variants, including one in *ITGB2*, may have enabled Japanese populations to survive long periods of low food supply by lowering their metabolic rate. But a genetic variant that is beneficial in times of famine becomes harmful in the presence of limitless supplies of high-calorie foods.

If you are a Japanese American suffering from or worried about becoming overweight or obese, knowing that you possess this genetic variant can be incredibly empowering. Armed with this information, you'll know that by following a more traditional Japanese diet or simply by cutting down on your consumption of animal fat, saturated fats, and sugars, you'll be able to stay slim.

Whether a diet of *any* type will help you lose weight is also determined by your genetic code. Although it is commonly thought that cutting down calories from any dietary source so that you consume fewer total calories than you use on a daily basis

Cutting down calories on a daily basis does not always lead to immediate weight loss.

should be sufficient to create immediate weight loss, genetics is showing us that this is not necessarily the case.

One study found that variants in the *ACSL5* gene may determine your chance of losing weight on a low-calorie diet, with one variant causing a person to be diet responsive and another diet resistant. Those with the diet-resistant variant appear to lose less weight than those with another variant. The diet-*responsive* genetic variant is present in about 15 percent of people, and the diet-*resistant* genetic variant is present in about 60 percent.

Both the *ACSL5* and *PPARG* genes play a significant role in delivering energy from fat cells to muscles. When a person is dieting, the body uses these genes to switch from using food as an energy source to breaking down fat cells so that the stored fat can be used for energy. In people who are diet responsive, a variant in the *ACSL5* gene increases the amount of body fat broken down, but in those who are diet resistant, another variant in the same gene causes the body to conserve fat instead of burning it. Variants in the *PPARG* gene have a similar effect.

The diet-resistant variants in these genes may have been necessary for survival in the past when food was often scarce, and this probably explains why they are more common than the diet-responsive variants in all populations. But now, in the 21st century in industrialized countries, the diet-resistant variants may be making it difficult for people to lose unwanted weight. The initial data also indicate that variants in the *ACSL5* and *PPARG* genes may have a synergistic effect, meaning that if you have a diet-resistant genetic variant in both these genes you'll have an even harder time losing weight through dieting.

TIPPING THE SCALE WITH A CUP O' JOE?
Caffeine, Sleep, and Obesity

HANDS DOWN, caffeine is the most popular drug in the world. More than 90 percent of Americans consume caffeine each day, and worldwide, tea is one of most commonly consumed beverages, second only to water. Consuming caffeine causes changes in your brain: First, it releases a hormone called dopamine,

continued

which elevates your mood and is responsible for it making you feel good, and, second, it blocks a brain-slowing and tiredness-inducing chemical called adenosine. But caffeine could also have some unwanted effects, such as disturbing your sleep and increasing your risk of obesity.

Some people have adverse reactions to caffeine without even knowing it. For example, while caffeine is known to negatively affect sleep, people who consume caffeine only in the morning or afternoon usually assume that their sleep at night won't be affected.

A preliminary study has found that caffeine's effect on sleep is dictated by variants within the *ADORA2A* gene, which codes for the adenosine receptor in the brain—the one that caffeine binds to, to increase your alertness. People with a variant in this gene are much more likely to experience prolonged effects from caffeine, and the study found that when they drink caffeine during the day, they are less able to sleep well at night. As a result, they may feel more tired and run-down in the long term. And the more tired they feel, the more caffeine they may drink, which can lead to a vicious cycle.

Chronic abnormalities with sleep patterns, such as not getting a good night's rest over a long period of time, is one of the many causes of obesity. Inadequate sleep produces disturbances in a person's circadian rhythm and hormone levels that, in turn, change the body's metabolic rate and increase the likelihood of gaining weight. Because of this, anything that leads to sleep disturbance may lead to obesity.

Yes, even the ability to keep weight off after you've successfully shed those pounds is predicated in part on your genetic code. For example, one study found that a variant in the *PPARG* gene decreased the body's ability to build new fat cells, thereby limiting the amount of weight a person was able to regain. As a result, individuals with this genetic variant (just over 20 percent of the study population) were much more likely to maintain their weight loss for more than a year than those who didn't have the variant. Knowing that you have this protective variant could certainly increase your motivation to lose weight in the first place.

VEGGIES: TASTY OR NASTY?
It's in Your Genes

IN 1932, Dr. Arthur Fox of the Jackson Laboratory in Wilmington, Delaware, was placing a powdery chemical called phenylthiocarbamide (PTC) in a bottle when some of its dust flew out into the air. Another laboratory worker who was nearby at the time complained of a bitter taste in his mouth, but Dr. Fox, who was actually holding the bottle, didn't taste anything. From this observation a debate ensued, leading to both of them tasting the PTC directly from the bottle (scientists have a long history of using themselves as guinea pigs). Once again the colleague stated that the PTC tasted intensely bitter while Dr. Fox was equally adamant that it had no taste whatsoever.

Dr. Fox tested this taste discrepancy in a larger population and found that some people, referred to as tasters, described a bitter taste while others, nontasters like Dr. Fox, didn't taste anything. Dr. Fox went on to discover that it was the molecular structure of PTC that caused some people to be tasters and that other chemicals with similar molecular structures also elicited the same effect. About 70 percent of people throughout the world are tasters.

In 2003, a researcher at the National Institutes of Health found that whether a person is a taster or nontaster of PTC results from variations in the *TAS2R38* gene, which codes for a receptor on your taste buds that's responsible for sensing bitter substances.

continued

But how does this relate to tasting the foods we eat? One possible explanation has to do with a chemical called goitrin, whose molecular structure is very similar to that of PTC. Goitrin is found in raw cruciferous vegetables—cabbage, broccoli, kale, Brussels sprouts, cauliflower, and collard greens. And for many people goitrin is bitter tasting.

Many genetic variants found to occur frequently within a population likely provide (or once provided) a survival advantage. Variants within the *TAS2R38* gene are no different. As it turns out, goitrin has a nasty side effect: It interferes with iodine metabolism, causing your thyroid gland to reduce its production of hormones, leading to hypothyroidism. But this occurs only if a large amount of goitrin is consumed on a frequent basis, meaning that you'd have to eat a whole lot of cruciferous vegetables. In the ancient past, people subsisted on whatever food was most accessible to them and, at times, only a single food source was available. If all they ate during those times were cruciferous vegetables, they'd be increasing their risk of developing hypothyroidism, a potentially serious disease. Researchers believe that people who were tasters were much less likely to eat large quantities of these vegetables because they tasted nasty; therefore, tasters were more apt to search out non-goitrin-containing foods. Thus the variants in the *TAS2R38* gene may have provided a survival advantage by protecting people from developing hypothyroidism.

Nowadays, these same genetic variants may dissuade people, especially children, from eating their vegetables. And because cruciferous vegetables have been associated with a lower risk of many diseases, finding ways to outsmart this gene and enable people to eat more of these vegetables on a regular basis is quite important. One way to do this, for example, is to recommend mixing the vegetables with other foods to mask their bitter taste.

While information about *why a* person likes or dislikes the taste of a particular food may not seem to be all that important, when it is integrated into a disease matrix it does become clinically useful. For example, if a person is found to be genetically predisposed to lung cancer and it is determined that eating cruciferous vegetables is likely to decrease that risk, knowing that he or she will probably not like the taste of these vegetables becomes valuable information. Instead of a physician's recommending that the person "eat more cruciferous vegetables," he or she might say "eating cruciferous vegetables is likely to decrease your risk of lung cancer but, based on your genetic makeup, you probably won't like the taste of those specific vegetables, so it may be best to mask their flavor by mixing them with other foods." In essence, this personalized information may be one of the best ways we have to get people to eat their vegetables.

Interestingly, smoking cigarettes can elicit the same bitter taste among tasters, and numerous studies have found that tasters are protected against nicotine dependency, while nontasters have a much higher likelihood of becoming addicted to smoking. This may be one reason some adolescents who experiment with smoking go on to become daily users while others say it tastes nasty and either don't smoke again or smoke only occasionally.

GENETICALLY TAILORED FITNESS

Recently, scientists have found that fitness and exercise, like weight loss, have genetic underpinnings. It is inherently much easier for some people to exercise and become fit than it is for others. In addition, the two basic aspects of what we call fitness—power and endurance—are exclusive of one another, so the person who has inherited an ability to gain muscle strength may not do so well in endurance sports, and those who find it easy to increase their endurance through aerobic exercise may have a much harder time with sports that require intense bursts of power. Even elite endurance athletes usually perform poorly in power events and vice versa.

Based on their genetic makeup, some individuals are inclined toward endurance events, which consist of sustaining low- to medium-intensity physical activity for long periods of time. Examples of endurance-related physical activities are long-distance running, rowing, long-distance biking, and cross-country skiing. Other individuals are genetically inclined toward power events, which involve high-intensity physical activity, usually for less than 20 minutes at a stretch. Examples of power-related physical activities include sprinting, swimming less than 600 meters, gymnastics, soccer, volleyball, wrestling, downhill skiing, tennis, boxing, and weightlifting.

Genetic variants can predispose you to excel at specific types of sports and exercises.

Variants in a number of genes predispose a person to be either an endurance or a power athlete. One of the most widely studied is the *ACTN3* gene. This gene is very active in a specific type of muscle made up of primarily fast-twitch fibers. In people with the endurance-based *ACTN3* variant, the muscle fibers exhibit a shift toward aerobic metabolism, which means they are able to use oxygen as a power source without tiring or cramping for long periods of time. The other variant in this gene predisposes individuals to power-based activities because it allows the muscle fibers to produce intense movements for short periods of time before the muscle has to rest.

The variant associated with power is often found in elite gymnasts, soccer players, judo masters, short-distance swimmers, speed skaters, and sprinters. The endurance variant is often found in marathon runners, long-distance cyclists, long-distance swimmers, cross-country skiers, and rowers.

Variants in the *ACE* and the *EPAS1* genes, which are involved in the energy metabolism of muscle cells, have also been associated with athletic performance. Both these genes have been studied in world-class athletes and Olympians, and just like the *ACTN3* gene, some variants were found much more often in elite endurance athletes and others in elite power athletes.

A comprehensive analysis of your genetic makeup that takes into account all these genes and their possible variants can be used to create a fitness program tailored specifically for you. Without that information, however, you might be attempting to follow a training program that is inconsistent with your athletic predisposition, and this may increase your fatigue and frustration, possibly leading you to give up on physical training altogether. A recent study published in the *Journal of Sports Medicine and Physical Fitness* showed that women experienced greater improvement from their fitness training when their program took into account their genetic profile. For trainers, this means being better able to motivate clients by providing them with a truly personalized program.

Comprehensive genetic analysis can now provide you with a fitness and athletic program tailored specifically for you.

Although some people may already have a good idea about whether they are more comfortable with endurance or power training, genetic screening can provide further guidance and objective confirmation. This can be very useful for fitness trainers who want to develop a personalized program that is tailored to their clients' genes. And it is also a way to inform parents and coaches of a child's athletic predisposition so that they are able to focus on sports that are best suited to the young athlete's genetic makeup from the start, potentially helping the child achieve higher levels of success.

LIVE LONG AND PERSPIRE

A 2008 study conducted by researchers at King's College in London found that the telomeres, which are the very end regions, or caps, of our chromosomes, are, in essence, our genetic clocks. When we are born, our telomeres are very long, but as we age, they become shorter and shorter. When the telomeres become too short, the chromosomes do not function correctly, and the cell becomes damaged. This sends a signal to the cell that it is approaching the end of its days, and the cell starts to prepare for its own death. This doesn't mean the person dies, just that, due to telomere shortening, the cell starts to grow old and withered. And if that cell then divides, its progeny cells will also be old and withered. Much of the destruction of the body that occurs with aging is attributable, on a genetic level, to decreasing telomere length. However, the clock does not tick at the same rate for everyone.

The study found that people who exercised regularly (40 minutes per day) had significantly longer telomeres than those who did not exercise. In fact, individuals who participated in heavy exercise on a regular basis had telomeres that were so much longer than those of individuals who did not exercise that the exercisers were, on a genetic level, up to 10 years younger!

Exercising on a regular basis actually slows down our genetic clock. And other studies have confirmed that the speed at which our genetic clock counts down depends on our level of physical activity. Because of this, two people, both of whom are chronologically 40 years old, may have two very different genetic ages.

It is interesting that these effects were primarily seen with leisure-based exercise, not work-based physical activity. This most likely occurs because leisure-time exercise has significant benefits not only for your body but also for your mind as it helps alleviate psychological stress—another variable (along with cigarette smoking and being overweight) that has clearly been shown to speed up a person's genetic clock.

A question I often encounter from my patients is, "Exactly how much exercise do I need to do to slow down my genetic clock?" Research indicates that, on average, you'll have to perform cardiovascular exercise (physical activity that increases your heart rate) for at least 20 minutes 5 days a week, or for 35 minutes 3 days a week.

The faster your genetic clock is counting down, the older you will feel, and this self-perceived early aging might then lead to a decreased will to change your lifestyle and start exercising, which creates a vicious cycle: The less active you are the older you feel, the older you feel the less active you are, and so on. The best way to avoid this downward spiral and beat your genetic clock is by incorporating regular exercise into your daily routine.

Using this information about the effect of physical activity on one's internal genetic clock, physicians and fitness trainers will be able to objectively evaluate the effectiveness of your exercise program over the long term. If you are truly achieving returns from your exercise, the results will be evident by genetically studying the length of your telomeres. If their length indicates that the results aren't adequate, you, your physician, and your trainer will know that you have to adjust your workout and step it up a notch.

While variations in some genes may predispose you to being an endurance or power athlete, others may actually predispose you to feeling fatigued with even a moderate amount of physical activity. One such variant, which occurs in the *AMPD1* gene, is found in more than 1 in 5 Caucasians, although it is very rare in other populations. The *AMPD1* gene produces an enzyme that is involved in delivering energy to your muscles. A variant in this gene causes about 75 percent *less* of the enzyme to be produced, thereby significantly decreasing the amount of energy that can be delivered to the muscles during exercise. As a result, individuals with this variant experience significant fatigue with exercise, which means they are more likely to decide that exercise just isn't for them and give it up altogether. However, research has shown that if a person is able to work through the initial fatigue and continue to work out on a regular basis, those feelings of fatigue will dissipate. As proof, the variant has even been found in world-class athletes and Olympians; thus it is possible to outsmart this variant in the *AMPD1* gene.

Among the many other genetic variants that help determine exercise performance and effectiveness, some affect the mitochondria (the parts of a cell that produce energy). Variants in *PPARD* and *PPARDC1A*, two genes that directly affect mitochondrial function, have been found to contain variants that decrease mitochondrial function and significantly reduce the benefits of exercising. Studies have shown that people with these variants experience poorer results from working out than people who don't have the variants. And while exercising primarily decreases a person's risk of diabetes by increasing the sensitivity of his or her body to insulin, individuals with these variants experience less of this beneficial response, meaning that exercise is also not as effective in decreasing their risk of diabetes.

Rare variations in a number of other genes also affect the mitochondria, some with drastic effects. For example, some cause extreme fatigue, muscle aches and pains, and exercise intolerance soon after commencing physical activity.

Just as there are variants that affect a person's endurance, there are also those that come into play during strength training. In one study, researchers at the University of Pennsylvania looked at a large number of people who had not engaged in any kind of resistance training during the previous year. The researchers put them on a 12-week resistance-training program and found that variants in the *IL15* and *IL15RA* genes were associated with the amount of muscle strength gained in response to the training. This most likely occurs because the *IL15* gene produces a protein that stimulates muscle growth and the *IL15RA* gene produces the receptor that detects the *IL15* protein; thus the two genes function in tandem. When both genes are functioning correctly, they send a signal to the muscle cells telling them to grow in size in response to resistance training.

THE DOPING GENE

EACH PERSON'S body is unique in the way it metabolizes almost any substance, be it food, medication, or illicit drugs. Because of this, a one-size-fits-all approach to determining how a substance will be processed by the body is usually not the most optimal approach.

An example of this is the use of anabolic steroids as performance-enhancing drugs. According to a survey conducted at the 1984 Olympics, nearly 70 percent of Olympic athletes had taken anabolic steroids at some point in their life. In 2006, almost half of all positive drug tests in athletes were due to anabolic steroids, a practice referred to as "doping." Notably, Arnold Schwarzenegger has admitted to steroid use during his bodybuilding career, and Floyd Landis was stripped of his title in 2007 when he tested positive for anabolic steroids after winning the Tour de France, although he still maintains his innocence.

The use of anabolic steroids is now banned in almost all sports around the world, and because of this, drug screening of athletes has become standard practice. The drug screen usually consists of testing a person's urine for signs of doping because some amount of the steroids taken is usually excreted in the urine. While the same cutoff value is used for examining each person's urine during this test, researchers have found that using one standard cutoff may cause some people to test positive even when they aren't doping and others to show up as clean even when they've been using. This occurs because of variations in the *UGT2B17* gene, which is involved in transporting both naturally produced testosterone and anabolic steroids from a person's blood to their urine.

Unfortunately, the doping test does not take into account the fact that genetic differences will cause some innocent people to fall above the cutoff (known as false positives) while

continued

some dopers fall below the cutoff (known as false negatives). The researchers found that using a cutoff value that was genetically tailored to the specific variants in a person's *UGT2B17* gene eliminated all false positives and all false negatives. Therefore, genetically tailoring the cutoff value greatly increases the accuracy of the test.

WILL WEIGHT-LOSS MEDICATIONS WORK FOR YOU?

The analysis of your genetic makeup not only can reveal that you are at increased risk for obesity and help determine what kind of diet and fitness program will be best for you but can also determine the effectiveness of weight-loss medications. Appetite suppressants, such as sibutramine (Meridia and Reductil), may help you lose weight, but they are effective for as few as 20 percent of people who try them. Therefore, 80 percent of people who take these medications will not experience significant weight loss, meaning a waste of money on something that isn't working. Preliminary studies have found that variants in the *GNB3* gene can be used to predict the effectiveness of sibutramine. For people with one variant, the medication is effective, but for people with another variant, it is not.

WILL WEIGHT-LOSS SURGERY WORK FOR YOU?

Weight-loss surgery (also referred to as bariatric surgery), which either decreases the size of the stomach or alters the way food passes through the digestive tract, leads to an average weight loss of 100 pounds or more. As with diets, exercise, and medications, however, some people unfortunately gain back a significant amount of the weight, and your genetic code can help predict whether that will happen to you.

Researchers have compared the outcomes of weight-loss surgery in people who have MC4R variants that predispose them to obesity with the outcomes of those who did not. The study participants all underwent weight-reduction surgery, and it was found that those who had MC4R variants lost less weight, had more adverse reactions, experienced a much higher complication rate, and had poorer overall outcomes than those who did not have any variants. It appears that people with MC4R variants are much more likely to continue eating excessively even after weight reduction surgery, which leads to not only regaining the weight but also serious complications because overconsumption is particularly dangerous once the size of the stomach has been surgically reduced.

Variants in a number of other genes are also being studied to determine their power to predict the effectiveness of weight-loss surgery. Bariatric surgery for weight reduction is always serious and can be risky, so knowing in advance whether you are likely to maintain the desired result means that you and your physician can make an informed decision.

TIPPING THE SCALE IN OUR FAVOR WITH GENETIC SCREENING

Many of the answers to achieving your personal weight-loss and fitness goals are within your genetic makeup. Now comprehensive genetic screening can provide you with access to that information, allowing you to tailor weight loss, nutrition, and fitness to your specific genes, thus empowering you to take control of your health and wellness.

NUTRITION, FITNESS, AND WEIGHT-LOSS PANELS

Panels increase the efficiency of genetic screening because they access all pertinent genes, traits, and diseases at one time and avoid nonrelevant information. What follows is an example of a panel I created for weight loss and nutrition.

Weight-Loss and Nutrition Panel

- Obesity and leanness
    - Includes BMI, waist circumference, and fat accumulation in specific areas of the body
- Exercise tolerance and athletic predisposition
    - Includes a genetically tailored workout regimen that will most efficiently enable weight loss
- Effects of specific diets on weight and obesity
    - Includes genetically tailored diets that are associated with increased weight loss
- Effectiveness of medications used for weight reduction
- Effectiveness of surgical procedures for weight reduction
- Taste perception and preference
- Caffeine metabolism
    - Including whether caffeine (consumed at any time, even in the morning) is likely to affect sleep quality at night
- Association of the consumption of specific foods with colorectal cancer risk
- Effect of specific diets on bone mineral density and osteoporosis risk
- Effect of specific diets on cholesterol levels
- Effect of specific diets on blood pressure
- Effect of specific foods and beverages on cardiovascular health and risk of heart attacks
- Metabolism of vitamins and nutritional supplements

For additional panels related to weight loss, nutrition, and fitness, please visit www.OutsmartYourGenes.com/Panels.

6

Prospective Parents
Getting Pregnant and Protecting Your Future Children

MISCONCEPTION: There's no way to predict what diseases and traits my *future* children will have because they haven't even been conceived yet.

FACT: Through a new approach called Pythia, we can now analyze the genetic makeup of two potential parents to determine which diseases and traits their future offspring are likely to have.

PROTECTING OUR MOST PRECIOUS RESOURCE

All parents want to do everything possible to protect their children and maximize their children's chances of happiness and physical well-being.

Advances in genetic screening now empower us to protect our children—as early as during gestation and even *before* they are conceived. Both rare and common diseases can now be screened for at any of those stages. When this type of genetic screening is combined

Predictive medicine and comprehensive genetic screening can now be used to protect our future offspring from disease.

with counseling by a healthcare professional, the results can be used to proactively protect our future offspring from disease.

There is already an ethical debate surrounding the availability and use of genetic information in relation to family planning, and that debate is certain to grow as the technology progresses and becomes more widely available. Entering into that debate is, however, beyond the purpose of this book. My job here is to explain what is available and how it can be used so that each person can make the most informed decision possible based on what is right for him or her.

FAMILY PLANNING IN THE 21ST CENTURY

Family planning is the only way to truly prevent genetic variants that are associated with disease from being passed on to our future children. While the term *family planning* is often used in the context of contraception, it actually refers to any planning that goes into having children and can take place prior to, or even after, becoming pregnant. Therefore, genetic screening of prospective parents as well as of pregnant women falls within the context of family planning.

The family planning options discussed in the following sections are important to consider if you have a disease, if you have a family history of a disease, or if genetic screening shows that you are at risk of passing on a disease to your offspring. They all pertain to both common and rare diseases, and while some of the technologies are still being perfected, they are all available right now.

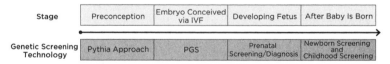

Stage	Preconception	Embryo Conceived via IVF	Developing Fetus	After Baby Is Born
Genetic Screening Technology	Pythia Approach	PGS	Prenatal Screening/Diagnosis	Newborn Screening and Childhood Screening

PREIMPLANTATION GENETIC SCREENING

Preimplantation genetic screening (PGS), also referred to as embryo screening, is a combination of genetic screening and in vitro fertilization (IVF). IVF is the procedure by which eggs are fertilized outside of the body in a laboratory, such as in a test tube or petri dish. About five days later, while the embryos are still in the laboratory, PGS is performed. The procedure involves taking a single cell from each of the embryos and conducting genetic screening to determine whether any of them has a debilitating or life-threatening disease. Based on the results of the screening, the physician is then able to implant only those embryos that do not contain the disease, thereby ensuring that the baby will be free of the disease in question.

Because PGS gives parents the option to determine which embryo(s) to implant, the resulting offspring are sometimes referred to pejoratively as *designer babies*, meaning they were designed by the parents to be a specific way, such as free of a specific disease. This type of selectivity puts PGS at the center of a polarized ethical debate: Some people believe it should be banned outright and others believe that it is a positive step.

In the 1990s, PGS was used primarily to screen for rare diseases like Tay-Sachs that caused suffering and death to a newborn or child. Then, around 2005, use of the technology was expanded to enable parents to ensure that their baby would be free from other diseases, such as childhood blindness. At this point, the technology evolved from preventing the inheritance of fatal diseases to offering parents the ability to ensure that their children would be free of severe disabilities. In 2009, use of the technology allowed for the first breast cancer–free baby. This child's parents had a significant family history of deaths due to breast cancer because of a variant in the *BRCA1* gene; they used PGS to choose embryos that did not have that genetic variant. For the first time, it was possible to substantially decrease the risk of a common disease that afflicts adults from being passed on to the next generation.

PGS can be used to decrease the risk of parents passing on both rare and common diseases to their children.

Critics stated that because common diseases aren't fully determined by genetics and don't manifest until adulthood, it isn't appropriate to choose embryos based on such characteristics. However, those who have seen their mother, grandmother, sisters, aunts, and cousins die of breast cancer might well argue that the disease had already killed too many members of their family and that it was their *right* to try to protect their children from the same fate.

The real controversy surrounding PGS, however, is not about its ability to stop the inheritance of diseases that cause tremendous suffering and death but about the fact that it can also be used to select for specific traits, such as hair and eye color, athletic performance, and intelligence. In a study published in 2009, a genetic counselor at New York University's School of Medicine surveyed close to 1,000 people who sought genetic counseling services. The study found that the majority of people would undergo genetic screening to prevent a serious disease from affecting their future children whereas only about 10 percent would screen for superior athletic ability, increased height, or greater intelligence.

Although the use of PGS for cosmetic purposes will certainly continue to be heavily debated, my primary concern, and the focus of this book, is not on trait selection but on disease prevention. In this regard, PGS is an incredibly powerful tool because it offers parents a true opportunity to ensure that their child will be free of a serious disease. PGS also potentially prevents parents from having to make a very difficult decision about whether to continue a pregnancy once a developing fetus has been diagnosed by amniocentesis, for example, to be affected by a debilitating or life-threatening disease.

As of now PGS with IVF is an expensive procedure that is still considered experimental by health insurance companies. Thus prospective parents have to incur great expense to take advantage of these technologies until they become more widely used and less expensive.

INVASIVE PRENATAL SCREENING

While PGS involves genetic screening before the transfer of the embryo into the uterus, another type of screening can be done on the developing fetus as it is growing inside the womb. This type of genetic screening is called prenatal screening and can be conducted in one of two ways: the traditional, *invasive* methods such as amniocentesis and chorionic villus sampling (CVS) or the newer, *noninvasive* methods referred to as noninvasive prenatal screening.

Both CVS (usually performed between the 10th and 13th week of pregnancy) and amniocentesis (performed between the 14th and 20th week) have been in use for decades and involve inserting a needle through a pregnant woman's abdomen directly into the womb and extracting cells that contain the genetic makeup of the developing fetus, which are then used for testing. Because a needle is being inserted into the womb, however, both of these procedures are considered invasive and carry risks. CVS has about a 1 percent chance of causing a miscarriage, and the risk with an amniocentesis is between 0.05 and 0.5 percent. Because of this risk, almost all of the diseases that are tested for using these invasive techniques are rare diseases such as Down syndrome and Tay-Sachs disease, which are either debilitating or deadly. These techniques are not used to screen a developing fetus assumed to be healthy but rather to make or confirm the diagnosis of a suspected disease because it is known to run in the family or because other tests, such as an ultrasound or blood test conducted during pregnancy, indicate a potential abnormality.

CVS and amniocentesis are usually covered by many health insurance companies when other screening tests have indicated a problem or if the pregnancy is considered to be high risk, such as if the woman is older than 35.

NONINVASIVE PRENATAL SCREENING

A number of revolutionary technologies are now paving the way for *noninvasive* prenatal screening. These procedures obtain genetic material from the fetus simply by taking blood from a vein in the pregnant mother's arm just as one would in any routine blood draw.

This technology, which will soon be available, is possible because the mother's and the fetus's blood supplies are connected. For example, when a pregnant woman eats food, that food is digested and absorbed into her blood. Some of the nutrients are used by her body

but some are also passed on to the fetus; conversely, other substances are passed from fetus to mother. Scientists have discovered
that both fetal cells and even
free-floating strands of fetal
DNA circulate in the mother's
blood. These can be extracted
from the woman's blood, purified, and used to determine the
fetus's genetic makeup.

Noninvasive prenatal screening allows doctors to study the genetic makeup of a developing fetus without any risk of miscarriage.

In the near future, we'll be
able to use this technology to
perform noninvasive genetic testing on the developing fetus to screen
for serious diseases. Just as parents receive copies of their ultrasound
pictures during pregnancy, I believe it will soon become commonplace
for parents to also receive the genetic makeup of their unborn child.

BREAST MILK
It Does a Mind Good—Or Does It?

IT'S BEEN known for a long time that, in addition to having a
stronger immune system and possibly a lowered risk of sudden
infant death syndrome (SIDS), babies who are breastfed have,
on average, a higher IQ than those who aren't. This is because
there are particular fatty acids in breast milk that are not in
formula or cow's milk. These fatty acids are found in high concentrations in an infant's brain and, in essence, act as brain
food. When there's more food available, such as from breastfeeding, the developing brain becomes smarter. This isn't to say
that non-breastfed infants can't grow up to be really smart or
that all breastfed infants will grow up to be highly intelligent;
what it does mean is that breastfed babies have a greater likelihood of having a high IQ. What we didn't know until recently,
however, was whether breastfeeding was interacting in any way
with the baby's genes.

In 2007, researchers at Duke University conducted an impressive study of more than 3,000 children to answer that question. After correcting for other potential variants such as socioeconomic status and parents' IQ, the results showed that a genetic variant in the *FADS2* gene was actually responsible for determining whether breastfeeding would increase an infant's IQ. Those who have the genetic variant will have an increase of approximately 7 IQ points, whereas infants without the variant won't experience any IQ boost from breastfeeding. This is significant because designations on the IQ scale (such as superior intelligence or average intelligence) are generally separated by about only 10 points.

The *FADS2* gene is involved in how a baby's body processes those special fatty acids found in breast milk. The variant that increases IQ with breastfeeding allows for a much more efficient processing of those fatty acids so the brain is exposed to more of the beneficial substance.

Having this information available during pregnancy or just after birth can be extremely useful. Although it may be ideal for most mothers to breastfeed, many are in situations that make the decision not so simple. For example, some women take medications that prevent them from breastfeeding, and others experience pain, poor milk production, and other difficulties. Now, with this link between the variant in the *FADS2* gene and an increase in IQ, we can provide these women with information that will allow them to make a more informed decision.

GENETIC SCREENING WITH ASSISTED REPRODUCTIVE TECHNOLOGIES

Assisted reproductive technologies (ARTs) are procedures that help couples overcome difficulty conceiving. One of the most common forms of ART is in vitro fertilization. Genetic screening can be conducted at a number of different points throughout IVF: before conception (such as screening the parents), after conception but before the embryos are transferred to the uterus with PGS, or after implantation has occurred.

The purpose of IVF is to allow individuals who have difficulty conceiving, or who may be unable to conceive, the opportunity to have children who are biologically linked to them in some way (as opposed to adopting). If there is an issue with either the sperm or the eggs, the couple may choose to use a donor. Egg donors may be found through a service, and sperm may be obtained from a sperm bank. No matter where the egg or sperm come from, however, genetic screening can help ensure that the child conceived through ART is healthy and free of serious disease.

Comprehensive genetic screening can now be used to detect genetic diseases in sperm and egg donors.

Often when someone purchases eggs or sperm, he or she is provided with some information about the donor, such as education level, the medical history of the donor and the donor's family, and specific physical traits. Although some sperm banks and egg donation services may conduct genetic screening for a few rare diseases, *comprehensive* genetic screening of sperm and egg donors is a relatively new concept that will most likely become a key differentiating factor in the donor industry. For example, in 2009, a lawsuit was filed by a 13-year-old child born with fragile X syndrome, which is a disease that causes mental disability. The disease had been inherited from sperm purchased from a sperm bank that had not screened the donor for this disease. The child claimed that because the sperm had contained a disease, this was equivalent to the sperm bank's having provided a defective product and sued under the product liability law. Because genetic screening technologies are so much more powerful than they were when this child was born, comprehensive genetic screening is now much more feasible, and hopefully diseases such as this will be detected so that they are no longer passed along to unwitting recipients.

ADOPTION

If either you or your spouse is found to be a carrier of or affected by a very serious disease and if, for any reason, you'd rather not pursue any of the other methods available, adoption is another family planning option to consider. A recent survey conducted by the U.S. National Center for Health Statistics found that there are about 120,000 adoptions in the United States each year, and as of 2002, approximately 2.5 percent (1.6 million) of all children currently living in the country were adopted. The option you choose depends on your own beliefs and what you think is best for you and your family. The important point is that options exist, so you *do* have a choice when it comes to family planning.

AU NATUREL

Just because comprehensive genetic screening is available, that doesn't necessarily mean it's right for you. It is your personal choice, and while healthcare professionals can provide you with information and help answer your questions about the various options, you ultimately have full autonomy when it comes to deciding whether genetic screening is right for you. Furthermore, if you do go ahead with the screening, and your future child is found to be at increased risk of a disease, proceeding without any intervention is also always an option.

GENETIC SCREENING OF PROSPECTIVE PARENTS

FERTILITY ISSUES

Approximately 1 in every 7 couples throughout the world experiences some form of fertility problem. These problems can range from difficulty conceiving to difficulty carrying a pregnancy to term. Although some difficulties conceiving (such as those that result from chronic infections or obesity) are nongenetic, many of the causes are genetic.

If a genetic issue is identified, a healthcare professional is then able to use this information to guide the prospective parents through the treatment and family planning options that are best for them.

Knowing if there are genetic reasons for fertility problems helps doctors to determine the best options for their patients.

One of the most common types of genetic variants associated with fertility issues are those that affect the viscosity and clotting properties of the mother's blood. Optimal blood flow is essential for pregnancy, and anything that interferes with the blood flow to the womb, such as if the blood is prone to increased clotting, can cause fertility problems. Because of this, clotting disorders account for up to 15 percent of recurrent miscarriages. As an example, variants in the *MTHFR* gene, which can cause blood clotting abnormalities, have been associated with both difficulty conceiving and recurrent miscarriages. However, variants within this gene can be outsmarted with a combination of a medication and two different B vitamins, allowing many of these women to have a normal pregnancy. A study published in 2006 investigated recurrent pregnancy loss in more than 500 women and found that it was necessary to conduct genetic screening on at least seven different genes, including *MTHFR*, to identify all the potential variants a woman might have that are known to be associated with blood clotting issues and infertility.

In addition, there are many other reproductive issues with a genetic component that might cause a woman to have difficulty conceiving and/or bringing a pregnancy successfully to term. Comprehensive genetic testing and analysis can screen for all these potential problems at once so that, if a genetic component is discovered, treatment or other family planning options can begin sooner rather than later.

In 50 percent of cases, it is the man who is the cause of a couple's infertility. The first step in the investigation of male infertility is always semen analysis, and many times the cause is either a very low sperm count or no sperm at all. Although genetic analysis isn't necessary to determine an abnormal sperm count, many times the reason for the abnormality is genetic in origin, and screening can identify those causes.

In the end, knowing why a couple has been unable to conceive will allow them to make better choices: either to correct the problem or, if that isn't possible, to determine the best alternative way for them to become parents.

WHO'S YOUR DADDY?

ONE OF the fastest-growing segments of the genetic testing industry is testing for paternity. An Internet search for "paternity DNA test" pulls up an impressive number of websites. The cost can be anywhere from $99 to a few hundred dollars, and some companies use CLIA-certified labs.

While paternity testing is by far more popular, maternity testing is also available. In fact, the first time genetic fingerprinting was ever used was in a case of questionable maternity. Genetic fingerprinting, which is similar to traditional fingerprinting, uses a person's unique genetic makeup to determine and confirm identity. Since we share 50 percent of our DNA with our mother and 50 percent with our father, using a person's genetic makeup to confirm maternity and paternity is relatively straightforward.

Genetic fingerprinting was invented by Sir Alec Jeffreys at the University of Leicester in England in 1985. After Jeffreys published his findings in *Nature*, the story was picked up by the media as an incredible breakthrough. Soon after that, Jeffreys was asked to help settle a two-year-old immigration case involving a Ghanaian boy who was stopped while attempting to enter Great Britain. The boy's mother, who lived in Britain, claimed the boy as her own, but immigration officials were skeptical. Jeffreys, using his newly invented genetic fingerprinting technique, was able to conclude not only that the woman was the boy's true biological mother but also that her other children were his siblings. As a result, the child was reunited with his mother.

ORPHAN DISEASES

Many, if not most, of the rare genetic diseases are considered statistically improbable, affecting fewer than 5 in every 10,000 people or fewer than a total of 200,000 people in the United States at any given time. Rare as they may be, however, such diseases still have a tremendous impact on our society. There are currently thousands of these orphan diseases, the majority of which are debilitating and life altering; many have a direct and detrimental effect on the fetus, newborn, or child. Some examples are Tay-Sachs disease, sickle-cell anemia, cystic fibrosis, muscular dystrophy, Niemann-Pick disease, and Bardet-Biedl syndrome.

Orphan diseases have special meaning to me not only because I have one myself, as I discussed in the introduction, but also because the first time I worked in a genetics laboratory I focused on an orphan disease called congenital erythropoietic porphyria (CEP). This is a rare, recessive disease that is caused by variants in the UROS gene, which normally plays an important role in converting a harmful substance in our blood into a useful one. When variants in the gene cause it to malfunction, the harmful substance builds up, causing a variety of symptoms, the most notable of which is to make the skin extremely photosensitive so that when people with CEP are exposed to sunlight, they blister and scar.

As discussed, people with a single copy of a genetic variant associated with a recessive disease are often called "carriers." Carriers almost never experience symptoms and usually don't even know they are carriers. Because of this, genetic variants that cause recessive diseases can unknowingly be passed from one generation to the next.

Because carriers don't actually have the disease, most people aren't even aware they have a genetic variant associated with a rare disease.

Even though they are rare, many people know of at least one family that has been affected by an orphan disease, which is why so many prospective parents fear them. With predictive medicine, we can now start to do something to mitigate that fear *before* a child is ever conceived by analyzing the genetic makeup of both prospective parents to see if either of them is a carrier. In my opinion, absolutely *no* disease is rare enough to be considered unimportant or to be excluded from a genetic screen for prospective parents.

ABE LINCOLN'S SECOND ASSASSIN
Was His Death MEN2B?

IT MAY surprise you to hear that Abraham Lincoln had a second assassin-in-waiting: cancer. If he hadn't been killed by John Wilkes Booth in 1865, it's likely that he would have died *within a year* from a rare disease called multiple endocrine neoplasia type 2B (MEN2B), which is caused by a genetic variant within the *RET* gene.

Lincoln was 6 foot 4 inches tall and quite thin throughout his life. Because of his notable appearance, some medical historians had previously thought that he might have had a rare genetic disorder called Marfan's syndrome. However, further research found that his medical history was inconsistent with the other key symptoms of that disorder; therefore, other theories were proposed.

John Sotos, a cardiologist and medical adviser for the television show *House*, conducted an extremely rigorous examination of Lincoln and his family's medical history. He found that all of the available information pointed to the idea that the president was afflicted with MEN2B.

Like Marfan's syndrome, MEN2B is characterized by a very tall, thin body type, but other symptoms include lumps created by an overgrowth of nerve cells usually visible around the lips, asymmetry of the face, constipation and other intestinal issues, abnormalities of muscle tone that can cause, among other things, loose-jointedness and droopy eyelids, and the development of life-threatening cancer at some point in the person's life. Lincoln

continued

had every one of these symptoms, and although we don't know whether he had cancer, he was severely emaciated at the time he was assassinated. This emaciation, along with a number of other symptoms that manifested shortly before his death— headaches, fainting spells, bouts of sweating, cold hands and feet, and fatigue—are symptoms associated with the types of deadly cancers caused by MEN2B.

Take out a $5 bill, and not only will you see his droopy eyelids but you may also be able to discern the outline of a characteristic lump on his lower right lip. Photographs of Lincoln taken during the final months of his life also provide other clues, including a gaunt appearance, prominent cheekbones, and sunken eyes. Some historians have attributed his wasted appearance to the hardships of the Civil War, but just a year earlier, when the war was already winding down, photographs show him to be considerably fuller of face and apparently healthier.

MEN2B is a very rare, inherited, dominant disease, which means that one of Lincoln's parents, 50 percent of his siblings, and 50 percent of his children were also likely to have been affected. Dr. Sotos found that Lincoln's family history was fully consistent with this: His mother had the same tall, thin appearance, and three of his sons had bumps around their lips. His mother, brother, and three of his four children (the ones with the bumps around their lips) also died young. Lincoln's fourth son, the only one without any of these symptoms, lived to the age of 82. Because Lincoln has no direct descendants living today, Dr. Sotos is currently trying to gain permission from museums that own materials containing the president's DNA to conduct genetic testing to confirm this diagnosis.

While the capabilities of medicine in Lincoln's time could not have saved his life even if his disease had been diagnosed, today people with MEN2B can live much longer, healthier lives so long as the disease is diagnosed early enough to monitor for cancer and other complications.

In the past, because of the cost involved, doctors rarely screened for orphan diseases, which meant that carriers didn't discover their status until their fetus, newborn, or child was diagnosed with a rare disease. Now that genetic screening technologies are more powerful and more cost effective, it is becoming possible to screen for practically *all* orphan diseases at once.

It is now becoming possible to screen for almost all orphan diseases at once.

By alerting people to the fact that they carry a disease, we are empowering them to make *informed* family planning decisions, such as having their spouse or potential parenting partner screened as well so that everyone will at least know the odds of their children inheriting a disease *before* the child is conceived.

COMMON DISEASES

Another benefit of genetic screening of prospective parents is that they can also learn about their risk of passing along common diseases to their future children. Learning about your own risk of common diseases empowers you not only to attempt to prevent their onset for yourself but also to understand exactly what you may be passing on to your children. Instead of just hoping that better treatments will be found by the time our future children are faced with such diseases, we can now defeat *their* future diseases by outsmarting *our* genes today.

PRECONCEPTION GENETIC SCREENING: THE PYTHIA APPROACH

For the past several years, I've been working on a new genetic screening technology called Offspring Projections through the Combined Analysis of Different Individuals (OP-CADI). OP-CADI can be used to predict the probabilities of diseases and traits in children who haven't even been conceived through the combined analysis of each potential parent's genetic makeup. Because the official name of this technology is quite long, my team has taken to calling it the Pythia Approach (named after a powerful oracle priestess from Greek mythology).

The Pythia Approach uses specially designed computer software to replicate the biological process of combining the genetic makeup from each parent, which is what occurs when a child is conceived, and then uses the resulting information to predict the range of possibilities for the child's genetic makeup. While we can't determine the exact risk of a common disease because the exact genetic makeup the offspring will

By analyzing the genetic makeup of two potential parents, we can predict the probabilities of diseases and traits that will affect their future children.

have is not yet known, by analyzing the prospective parents' genetic makeup we can predict the *range* of risk, from highest to lowest, that future children will have of getting the disease.

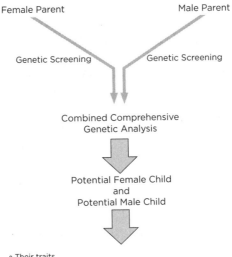

Female Parent   Male Parent

Genetic Screening   Genetic Screening

Combined Comprehensive
Genetic Analysis

Potential Female Child
and
Potential Male Child

○ Their traits
○ Common diseases they'll be at increased risk for
○ Common diseases they'll be at decreased risk for
○ Rare diseases they'll most likely have
○ Rare diseases they'll be carriers of (but won't actually have)
○ Rare diseases they will not have and will not be carriers of

The Pythia report is similar in concept to the easy-to-understand genetic report discussed in Chapter 3 but this specialized report contains the results from the Pythia Approach. The report contains two sections, one for male children and one for female children. This is because males inherit one Y chromosome from their father and one X chromosome from their mother while females inherit two X chromosomes (one from each parent). Because of this, the potential genetic makeup is different for males and females, meaning that the potential for inheriting genetic variants associated with some diseases will also be different.

What follows is one possible way the results of the Pythia Approach for a female child might be presented to you. Bear in mind that this example provides only a small sampling of the total information that would be included in the actual report.

For rare diseases

- *Cystic fibrosis:* 25 percent chance of having this disease, a 50 percent chance of being a carrier, and a 25 percent chance of being neither affected nor a carrier
- *Glucose-6-phosphate dehydrogenase deficiency:* a 50 percent chance of having this disease, a 50 percent chance of being a carrier, and a 0 percent chance of being neither affected nor a carrier
- *Alpha-1-antitrypsin deficiency:* a 50 percent chance of being a carrier but a 99.98 percent chance of not having this disease
- *Sudden death:* less than a 0.5 percent chance of sudden death due to deadly heart arrhythmias
- *Epidermolysis bullosa:* less than a 0.3 percent chance of having this disease
- *Retinitis pigmentosa:* less than a 0.1 percent chance of having this disease
- *Bardet-Biedl syndrome:* less than a 0.3 percent chance of being a carrier

- *Tay-Sachs disease:* less than a 0.03 percent chance of being a carrier
- *Sickle-cell anemia:* less than a 0.03 percent chance of being a carrier

For complex diseases:

- *Breast cancer:* 55 to 80 percent lifetime risk
- *Melanoma:* 10 to 17 percent lifetime risk
- *Colorectal cancer:* 12 to 21 percent lifetime risk
- *Obesity:* 4 to 12 percent risk in childhood; 14 to 22 percent risk in adulthood
- *Alzheimer's disease:* 4 to 13 percent lifetime risk
- *Heart attack:* 11 to 23 percent risk of early-onset; 21 to 36 percent lifetime risk

It is important to understand that this analysis, like medicine in general, is based on our present understanding of the genetic code. Although we can currently use this approach to predict the carrier status and risk of hundreds of diseases, the number of diseases will increase, and the probability ranges for the risk predictions will become narrower as we continue to learn more.

The Pythia Approach, which is in its final stages of development, is useful for anyone thinking about having children, including people contemplating the use of a sperm or egg donor, because it not only provides objective information about the genetic makeup of the donor but also about the risks for diseases that may result from the *combination* of both people's genes.

Examples of Reproduction-Related Genetic Screening Panels

Reproduction Panel/Sperm or Egg Donor Panel
(this panel is run on the prospective parents' genetic makeup)

- Cancer (all forms)

- Neurologic diseases (including muscular dystrophy, Alzheimer's disease, Parkinson's disease, Huntington's disease, Lou Gehrig's disease, and other neurologic diseases)
- Heart disease
- Stroke
- Causes of sudden death (including heart arrhythmias)
- Structural heart defects
- Risk of spina bifida
- Rare disease screen (including a comprehensive analysis of orphan diseases, metabolic diseases, and syndromes)
- Skeletal abnormalities and limb disfigurements
- Immune system diseases
- Mental retardation
- Pervasive developmental disorder (including autism, autism spectrum disorders, Asperger's syndrome, and Rett syndrome)
- Hearing impairment and deafness
- Visual impairment and blindness
- Longevity

Pregnancy Panel
(genetic screening for a pregnant woman using her genetic makeup)

- Child's risk of multiple sclerosis
- Child's risk of spina bifida
- Risk of preterm birth
- Preeclampsia, eclampsia, and/or hypertension during pregnancy
- Risk of large scar or of the incision not healing well after a C-section
- Blood clotting disorders, bleeding disorders, and other blood disorders
- Rare disease screen (including orphan diseases, metabolic diseases, and syndromes)
- Gestational diabetes

Embryo and Fetus Panel
(genetic screening for embryo or fetus using its genetic makeup)

- Personal genetic identifier for safety and security
- Blood group
- Rare disease screen (including a comprehensive analysis of orphan diseases, metabolic diseases, and syndromes)
- Skeletal abnormalities and limb disfigurement
- Mental retardation
- Pervasive developmental disorder (including autism, autism spectrum disorders, Asperger's syndrome, and Rett syndrome)
- Cancer (all forms)
- Neurologic diseases (including muscular dystrophy, Alzheimer's disease, Parkinson's disease, Huntington's disease, Lou Gehrig's disease, and other neurologic diseases)
- Immune system diseases
- Heart disease
- Structural heart defects
- Cardiac arrhythmias and causes of sudden death
- Effect of breastfeeding on IQ

Female Fertility Panel
(using the woman's genetic makeup)

- Female infertility
- Risk of miscarriages
- Premature ovarian failure
- Blood clotting disorders, bleeding disorders, and other blood disorders
- Disorders that affect reproductive ability (including primary and secondary sex characteristics and sex reversal)
- Hypogonadism (hormone abnormality)
- Polycystic ovary syndrome
- Thyroid abnormalities

- Rare disease screen (including a comprehensive analysis of orphan diseases, metabolic diseases, and syndromes)

Male Fertility Panel
(using the man's genetic makeup)

- Male infertility (including sperm production)
- Risk of erectile dysfunction (including high blood pressure, diabetes, and peripheral arterial disease)
- Effectiveness of erectile dysfunction medications
- Disorders that affect reproductive ability (including primary and secondary sex characteristics and sex reversal)
- Hypogonadism (hormone abnormality)
- Rare disease screen (including a comprehensive analysis of orphan diseases, metabolic diseases, and syndromes)

For additional panels related to family planning, pregnancy, and fertility, please visit www.OutsmartYourGenes.com/Panels.

7

Newborns and Children
Protecting Those Most Dear to Us

MISCONCEPTION: There's not much we can do to protect our children when they're young from diseases that may manifest when they're adults.

FACT: Prevention can start at any time, and genetic screening of children can help prevent and treat conditions that may affect them during childhood and later in life.

PERSONAL IDENTIFICATION AND SAFETY

For many years, newborns' footprints were routinely taken as a means of identification. Then, in the late 1980s and early 1990s, the American Academy of Pediatrics (AAP) began to advise against using newborn footprinting for identification because 9 out of 10 times it was done incorrectly, and the resulting prints were unidentifiable, making the practice useless. In addition, the AAP stated that there are much more accurate techniques that could be used, including genetic screening. Even though it isn't used as widely yet, genetic screening offers the most accurate and dependable form of personal identification, because it remains the same throughout life, it can't be faked, and it is detectable in almost any biological specimen, even if the person isn't present (for example, testing a hair or other sample left behind at a crime scene).

Newborns, children, and adults can find out their own unique genetic identifier by sequencing the letters of 25 to 50 different places in their genome. These places usually have nothing to do with disease risk and are relevant only as a personal identifier—because of this, the identifier doesn't provide anyone who may have access to it with any information about a person's risk of disease.

Genetic screening offers the most accurate and dependable form of personal identification.

GENETIC SCREENING OF NEWBORNS AND CHILDREN

In the previous chapter we discussed how genetic screening can be conducted at any time from preconception to birth. In this chapter we'll talk about the ways genetic screening can be made actionable at any time from the birth of a baby through the teenage years.

SCREENING FOR THE CAUSES OF SUDDEN INFANT DEATH SYNDROME (SIDS)

The number one cause of death in healthy infants after 1 month of age is SIDS. Just the idea that an apparently healthy infant can die suddenly for no apparent reason is terrifying to parents and devastating when it occurs. Although several causes—including the infant's sleeping on its stomach or being exposed to secondhand smoke—have been considered and are still being debated, the exact cause of SIDS has remained elusive. It is, however, becoming evident that a significant number of infants who die of SIDS have a disease called long QT syndrome, which is caused by variants in a number of genes. Long QT syndrome is an abnormality of the electrical conduction system of the heart that can spontaneously interrupt the normal rhythm of the heartbeat (called an arrhythmia), leading to sudden death. With comprehensive genetic screening we can start to identify infants who have the genetic variants known to cause long QT syndrome; empowered by this knowledge, we *can* institute preventive measures and protect these children against SIDS.

In 1998, the *New England Journal of Medicine* published a study by Dr. Peter Schwartz, one of the foremost authorities in the world on heart arrhythmias. For the study, Dr. Schwartz and his team performed a number of screening tests, including EKGs, on more than 30,000 infants. A number of these infants later died of SIDS, and by going back and assessing the EKGs that had been taken of them when they were alive, the researchers found that up to 50 percent had shown signs suggestive of long QT syndrome.

Then, in 2000, Dr. Schwartz published another article discussing the case of an infant who had appeared perfectly healthy until, 44 days after birth, his parents found him suddenly not breathing and without any pulse. They rushed him to the emergency room where doctors found that he had a life-threatening arrhythmia that occurs with long QT syndrome. Luckily, in that instance, they were able to shock his heart back to a normal rhythm, and the baby was given medication to stop the arrhythmia from recurring. The child continued to be given the medicine and, at 5 years old, was doing extremely well and no longer had any life-threatening symptoms.

At that point, Dr. Schwartz conducted genetic screening on the child and found that he had a variant in a gene that had previously been associated with long QT syndrome. Researchers then studied the boy's parents and found that neither of them carried the variant, meaning that it had arisen spontaneously in the child. The vast majority of these spontaneous new variants, referred to as occurring "de novo," are harmless and go unnoticed, but once in a while—as in this case—they can occur in an important gene and be associated with a serious illness.

As Dr. Schwartz points out in his article, if the infant had died, his death would almost certainly have been classified as SIDS and the underlying cause would never have been discovered.

If a physician knows that an infant's genetic makeup contains a harmful variant associated with long QT syndrome, the infant can be given a medication called a beta-blocker (the same medication the doctors gave the baby in the story above), which has been shown to decrease the risk of death from long QT syndrome from greater than 20 percent without medication to under 3 percent. And this is only one preventive measure. Others include frequent monitoring by a cardiologist and training parents to use a defibrillator (a machine that delivers a shock to the heart) at home in case of an emergency. Because so many cases of SIDS are attributable to long QT syndrome, detecting it and instituting preventive measures before symptoms manifest may significantly decrease the incidence of SIDS. As Dr. Schwartz concluded, "The practical importance of this concept lies in the fact that most deaths due to the long-QT syndrome *can be prevented*."

Genetic screening can identify serious conditions that may cause SIDS, thereby allowing preventive measures to be implemented before a life-threatening situation arises.

Conducting electrocardiogram (EKG) screening on all infants has been discouraged because it is considered an inaccurate and costly method of detecting long QT syndrome. With comprehensive genetic screening and the use of panels, however, we can screen for life-threatening conditions along with all other diseases and traits that may affect the infant, thus making the screening more cost effective while also increasing our chance of detecting any predisposition to disease that might exist.

SCREENING FOR AUTISM

Children with autism and a number of other related disorders such as Asperger's syndrome and Rett's syndrome exhibit a spectrum of developmental disabilities with varying degrees of severity that are often grouped under the umbrella term autism spectrum disorders (ASD).

Unfortunately, the number of children afflicted with autism has risen steadily over the last three decades, making it one of the most prevalent developmental disorders among children. Today, studies have found that 1 in every 100 children has some form of autism. And although the exact cause remains elusive, genetics is known to be responsible for about 90 percent of why a child gets autism, with nongenetic factors determining the rest.

About 25 percent of children who are diagnosed with autism by the age of 3 improve, and by the age of 7 many of them will be able to talk and enter school even though they continue to exhibit characteristics of autism. The remaining 75 percent, however, do not improve significantly and may need special care throughout their life. While medication and some forms of therapy can help with some of the symptoms, there is currently no cure for autism. However, early diagnosis and intervention have been shown to be life altering. Research has revealed that children who are diagnosed early and who receive therapeutic intervention at a young age are more likely to experience a decrease in autistic symptoms and a significant improvement in their social and functional abilities. Because of this, identifying children with autism through genetic screening provides extremely useful information.

Genetic screening that detects an increased risk of autism allows for earlier diagnosis and intervention.

An interesting feature of autism is that it is usually associated with a large head size; more than 80 percent of autistic individuals have a head circumference greater than the 50th percentile and almost 25 percent have a head circumference greater than the 98th percentile. Variants in the *PTEN* gene are found in more than 1 in 7 people who have both autism and an increased head circumference. The *PTEN* gene is responsible for *limiting* the growth of nerve cells in the brain. Variants can cause the gene to malfunction, which ultimately leads to an abnormal increase in the growth of nerve cells. Because of this, nerve cells in the brain become much thicker than normal and these abnormally large cells may be responsible for causing both autism and increased head circumference.

Variants in the *EN2* gene increase the risk of autism by more than 40 percent. However, unlike *PTEN*, the *EN2* variants appear not to cause autism directly but rather to increase the risk of the condition.

The *EN2* gene is essential for the proper development of the brain. When variants cause it to malfunction, the result is abnormal brain development. However, it's likely that variants in other genes, as well as nongenetic factors, interact with the *EN2* variant, causing autism. It is interesting that mice in which the *EN2* has been rendered nonfunctional exhibit autism-like behaviors, but even though both male and female mice had the nonfunctional *EN2* gene, many more males showed signs of the disease. The researchers proposed that the female mice might have other genetic variants that were protecting them from autism. This is an extremely useful finding because in humans the ratio of autism in males and females is about 4 to 1. In the past it was thought that males might be more prone to autism because of variants that affected only them, but it now appears possible that males and females are actually affected at the same rate but that females contain some other, as yet undiscovered, protective factor, such as another variant in their genetic code, that stops autism from manifesting. Finding this protective factor could potentially lead to a cure for autism.

One of the most contentious debates in the field of autism research centers around whether immunizations, such as the MMR vaccine, which immunizes children against measles, mumps, and rubella, increase the risk of autism. Before this immunization, getting infected by measles was thought to be as inevitable as death and taxes, but because of the vaccine the incidence of measles in the United States has decreased to less than 1 percent. More than 500 million doses of the MMR vaccine have been administered since it was first used in the 1970s, and this is estimated to have prevented more than 50 million cases of measles and close to 1.5 million deaths.

Clearly, the MMR and many other vaccines have been instrumental in ridding the world of serious infectious diseases, but some people now contend that childhood vaccines, such as the MMR vaccine, may be increasing the risk of, or outright causing, autism. The primary reason for their concern is the use of mercury-containing ingredients such as thimerosal, which is used as an antiseptic in some vaccines (although it was never used in the MMR vaccine). While the overall use of thimerosal has significantly decreased since 1999, it is still used in some vaccines. The primary argument is that the mercury (or some other substance in the vaccine) causes harm to the child's developing brain, leading to autism. However, a very large number of scientific studies have failed to detect any such association.

As a geneticist, I believe it may be necessary to do more tests that not only look at the association between vaccines and autism but *also* to include comprehensive genetic screening of all children in the studies to rule out the possibility that some children may have variants that predispose them to autism only when they are also exposed to certain vaccines. This is a reasonable hypothesis because specific enzymes are necessary to help the body process the numerous substances, including mercury, found in vaccines, and variants within the genes that produce these enzymes may change the way a child's body reacts to these substances. Instead of just looking at a few genes, the study should assess the entire genetic makeup of each child, and it should include thousands of children so that even rare variants and their effects will be detected. This approach has already identified many variants associated with hundreds of other diseases, and if a link between a genetic variant, a nongenetic factor (such as a vaccine or a specific ingredient within a vaccine), and autism is established, it could provide a way for us to predict risk before exposing specific children to whatever nongenetic factor is identified.

Large-scale studies are needed that investigate the association between a child's genetic makeup, autism, and various nongenetic factors such as certain vaccines.

PROTECTING YOUR CHILD FROM MULTIPLE SCLEROSIS

Multiple sclerosis (MS) is a chronic, usually progressive autoimmune disease affecting the nervous system. The name, which means "multiple scars," refers to the scarlike tissue that forms throughout the brain and spinal cord as a result of multiple attacks by the person's own immune system. The exact cause is still unknown, but many genetic variants and nongenetic factors have been associated with increased risk.

One of the most interesting findings is that people who live in countries that are very close to the equator have substantially reduced rates of MS compared to those who live in countries that are farther away, such as in North America and western and northern Europe. Many studies have looked at the association between the latitude at which one lives and one's risk of contracting MS, and the results have determined that those who live primarily near the equator and are exposed to a lot of sunlight during the first 15 years of life have a very low risk of MS; those who live in a colder climate and see less of the sun until that age, even if they then move closer to the equator, have a higher risk. Because of this, the risk associated with this nongenetic factor appears to be fully dependent on where one lives during the first 15 years of life.

One of the leading theories about the relationship between latitude and risk for MS is that people who live near the equator receive much more exposure to sunlight than those who live farther away. Since our bodies produce vitamin D only when the ultraviolet B (UVB) rays of the sun are absorbed through the skin, less sun exposure means our body is able to produce less vitamin D. The theory states that it is this decreased production of vitamin D that increases a person's risk of MS.

In 2009, researchers at the University of Oxford published a very insightful study that looked at the association between people's genetic makeup, their vitamin D levels, and their risk of MS. The study found that when the *HLA-DRB1* gene contains a specific variant, its functioning depends on vitamin D levels. When vitamin D levels are normal, the gene with this variant works normally, but when vitamin D levels are low, the gene malfunctions. If you don't have the variant, however, the *HLA-DRB1* gene works normally independent of vitamin D levels, so it's the variant that makes the gene's proper functioning depend directly on having adequate amounts of vitamin D in the body.

When low vitamin D levels occur in a child with a specific genetic variant, their risk of MS is significantly increased.

This same variant, which occurs in anywhere from 15 to 40 percent of people, had already been associated with a substantially increased risk of MS. Thus the *HLA-DRB1* variant provides a direct link between vitamin D levels and MS.

The *HLA-DRB1* gene is involved in teaching our immune cells what's foreign and should be attacked, such as bacteria or viruses, and what's part of our own body and should not be attacked, such as the neurons in our nervous system. If the immune cells go a bit haywire and become overreactive (as some do once in a while), the *HLA-DRB1* gene helps the body identify and destroy these harmful cells. This is extremely important during childhood, and even during fetal development, when the body is using *HLA-DRB1* to teach its immune system what it needs to know. If a child with a variant in the *HLA-DRB1* gene has a low level of vitamin D, the gene malfunctions, and the immune system can't be properly trained. This can lead to a buildup of rogue immune cells that can then go on to harm the nervous system, thereby causing multiple sclerosis.

Unfortunately, a very large proportion of many populations around the world have low vitamin D levels. Studies have shown that *1 out of every 2* Americans has vitamin D deficiency. And low levels of vitamin D have been reported throughout Europe and Asia, with a high proportion of the deficiency occurring in pregnant women and infants. But even in places like Hawaii and southern Arizona, where one would assume most people get enough sun exposure, about 50 percent of people are still vitamin D deficient.

The risk is even higher in people with darker skin for basically the same reason that these people are less likely to be sunburned. The darker skin stops a lot of the UVB rays from penetrating deep enough into the skin to stimulate the body's production of the vitamin. Therefore, darker skin helps protect against sunburn but also increases the risk of vitamin D deficiency. Because of this, almost 90 percent of African American children and over 75 percent of Hispanic American children have been found to have suboptimal levels of vitamin D.

As long as you know the importance of consistently checking your child's vitamin D level, your doctor will be able to identify and easily correct low levels if they are ever detected.

Vitamin D levels are easily determined through a simple blood test. If your children are found to be deficient, there are various steps you can take to ameliorate the problem. If you live in a sunny place, you can make sure that they spend 10 to 15 minutes outdoors every day without sunblock. Unfortunately, vitamin D deficiency is difficult to correct through diet, except by drinking cod liver oil. Personally, I'm not too fond of the taste of cod liver oil and I'd guess that your children aren't either, so one of the easiest ways to increase their dietary intake is to give them vitamin D3 supplements. And no, a multivitamin will probably *not* contain enough vitamin D to do the trick. Most multivitamins (even prenatal multivitamins) contain between 200 and 400 IU of vitamin D3, but adequate supplementation usually requires about 1,000 IUs per day. Your doctor will be able to tell you exactly how much supplementation you or your child needs if a vitamin D deficiency is detected.

Because many different variants and nongenetic factors are involved in the risk for multiple sclerosis, identifying this one variant and correcting any vitamin D deficiency may not definitively prevent the illness. But taking these steps will help substantially decrease your child's risk of MS and perhaps avoid it altogether.

THE WARRIOR GENE
Child Abuse, Antisocial Behavior, and Criminality

A VARIANT in the *MAOA* gene in males is associated with impulsive tendencies, antisocial behavior (defined as arrests for violent crimes and/or psychiatric symptoms such as wanting to inflict harm on other people), and a greater likelihood of engaging in physical violence. When scientists attempted to replicate this finding in mice by altering the *MAOA* gene in a similar way, they found that male mice also exhibited extremely aggressive behavior and went on the offensive in attacking other mice. The same violent tendencies were also observed in monkeys with a similar genetic variant in their *MAOA* gene. Because of its association with violence and aggressive tendencies in humans and other species, the media has dubbed *MAOA* the warrior gene.

Dr. Avshalom Caspi, currently an esteemed professor of psychology and neuroscience at Duke University's Institute for Genome Sciences and Policy, conducted a study in 2002 that identified a specific variant in the gene as being associated with a significantly increased risk of violence in men who had been emotionally, physically, or sexually abused as children.

Dr. Caspi investigated the 25-year histories of more than 1,000 white men and found that 85 percent of those who had the variant and who had been physically, sexually, or emotionally abused as children went on to develop some form of antisocial behavior as adults. Those who didn't have the *MAOA* variant but were abused as children did *not* have an increased risk of antisocial behavior, and neither did those who had the variant but were not abused. So it appears that for many individuals, both the variant and the child abuse are required to cause the development of antisocial behavior. In terms of violence alone, close to 45 percent of all violent offenders in the study had the variant and were abused as children, meaning that this genetic–environment interaction is also highly associated with violent tendencies.

The *MAOA* gene is involved in the control of chemical signals in the brain that have a direct impact on how the brain processes emotions and thought. The variant associated with antisocial behavior causes the gene to malfunction, leading to observable differences in how the male (but not the female) brain processes incoming information. The brains of these individuals appear to have an increased sensitivity to emotions and memories but a decreased ability to control impulses.

Childhood abuse has been known to cause significant changes—including changes in neurotransmitter levels—to the brain that persist throughout adulthood. When individuals who do *not* have this *MAOA* variant suffer abuse, their brain is more resilient because the gene is functioning normally. In essence, the *MAOA* gene helps correct the harmful changes to the brain induced by childhood abuse so that, given time, it can heal itself. But when the gene contains this specific variant, the brain is hypersusceptible to the changes in neurotransmitter levels and, therefore, much more susceptible to suffering significant, long-term harm, which may eventually manifest as violence, aggression, and antisocial behavior.

The reason that the variant in the *MAOA* gene affects only males is most likely because it is on the X chromosome. Males have only one X chromosome and therefore only one *MAOA* gene. If it contains this variant their only *MAOA* gene will not function correctly. Females, on the other hand, have two copies of the X chromosome and will only rarely have the *MAOA* variant on both copies. This means that females are much more likely to have at least one *MAOA* gene that's functioning normally and thus are generally protected from such behavioral problems.

continued

In 2009, Dr. Kevin Beaver, from Florida State University's College of Criminology and Criminal Justice, published a study of more than 2,000 Americans confirming that the same variant in the *MAOA* gene also increases the likelihood of males (but not females) joining a gang and using weapons. Looking at the risk of weapons use for people already in a gang, Dr. Beaver also found that those who possessed the *MAOA* variant were more than 200 times more likely to use weapons during a fight than gang members without the variant. As Dr. Beaver states, these findings seem to imply that even the differences in violent tendencies among gang members are determined in part by our DNA.

The findings linking the *MAOA* variant, childhood abuse, and antisocial behavior have been corroborated by a large number of studies. Because childhood abuse appears to be a necessary component in this harmful triad, child welfare services could potentially use genetic screening to identify children under their watch who are at highest risk and could take additional, proactive steps to limit their exposure to environments that might instigate antisocial behavior later in life. Pediatricians, social workers, and other healthcare providers could also monitor these children and their home environment for signs of abuse, and forensic profilers and correctional facilities could use the information to identify criminals at increased risk of violent behavior and weapons use.

DYSLEXIA—THE GENETIC COMPONENT

Dyslexia is a disorder that affects the way the brain understands and processes written and spoken language. It affects almost 80 percent of all people diagnosed with a learning disability, and, according to the International Dyslexia Association, as many as 1 in 5 school-age children and adults are believed to have some degree of the disorder.

While a plethora of theories have been proposed, no definitive cause of dyslexia has yet been identified. And while there may not be any outright cure, there are many techniques and interventions that allow people to overcome or compensate for the problem once it has been identified.

As with all the other traits and diseases discussed throughout this book, dyslexia and reading difficulties have a clear genetic basis, and multiple genetic variants have already been associated with both. Genetic causes are estimated to be responsible for 60 percent of all reading disorders, including dyslexia.

Genetic causes are estimated to be responsible for 60 percent of all reading disorders, including dyslexia.

Variants in the *DCDC2* gene, which helps nerve cells in the brain position themselves correctly during early brain development, have been associated with dyslexia. Researchers believe that these variants most likely interact with variants in the *KIAA0319* gene, which is also important for the initial positioning of the neurons responsible for processing language. When variants occur in both of these genes simultaneously, they interrupt the normal functioning of the genes, leading to a slightly abnormal wiring of the circuitry of the brain, which, in turn, leads to dyslexia.

As is true for most complex diseases and disorders, dyslexia is caused by a constellation of genetic variants in multiple genes that are interacting with various nongenetic factors. What genetic screening can provide is an indication of a propensity toward dyslexia, which will allow for an earlier, more definitive diagnosis and thus earlier interventions. Instituting proper interventions earlier in life has the potential to limit frustration and empower the child with the tools he or she needs to overcome the disability. As with obesity, infertility, and any number of other problems, learning that there is a specific genetic component at work not only alleviates guilt but also allows the person with the problem to focus on finding a solution.

IDENTIFYING ATTENTION-DEFICIT/HYPERACTIVITY DISORDER (ADHD)

ADHD affects up to 7 percent of all children. Approximately 90 percent of the cause is determined by genetics and 10 percent by nongenetic factors, such as whether the mother smoked when she was pregnant.

A number of variants, primarily in genes associated with neurotransmitters in the brain, have been found to increase the risk of ADHD. Although the effects of each of these variants is moderate, when taken together, they may have a significant effect on neurotransmitter levels, causing abnormal brain function and the symptoms associated with ADHD. Variants in the *DRD4* gene, for example, affect the neurotransmitter dopamine and appear to be responsible for increasing the risk of impulsivity and excitability. Other variants in the *SLC6A4* gene affect the neurotransmitter serotonin and are likely to be responsible for increasing the risk of inattentiveness and fidgetiness. And even more variants in the *SNAP25* gene may be responsible for increasing the risk of hyperactive behavior. These, in conjunction with other variants in similar genes, all affect the way the brain works. Additional studies are under way to better understand how these, and any additional variants, may work in synergy to cause ADHD, and we hope to have truly actionable data within the next few years.

SCREENING TO PREVENT OBESITY AND DIABETES

In Chapter 5 we discussed the specific variants that increase the risk of childhood obesity. Knowledge of a child's predisposition to obesity will help that child and his or her family to overcome misplaced feelings of guilt and increase self-esteem. Even more important, however, genetic screening can be used to discover the variants that will help the child outsmart his or her genes and lose excess weight.

One of the primary causes of diabetes is obesity, and many of the same variants that increase children's risk of obesity also increase their risk of diabetes, but only if they become overweight or obese. Thus overcoming the genes that predispose children to obesity will often allow them to avoid type 2 diabetes as well.

PREDICTING AND PREVENTING ALZHEIMER'S DISEASE

You may be surprised to see Alzheimer's disease discussed in conjunction with genetic screening for children, but prevention of Alzheimer's disease can start at any age, and from the perspective of predictive medicine, the earlier you are aware of an increased risk of this disease, the sooner prevention can be implemented.

Head trauma, especially repeated head trauma that comes from playing contact sports such as boxing, hockey, and football, has been associated with an increased risk of Alzheimer's disease. A considerable portion of the risk of Alzheimer's after head trauma is due to a variant in the *APOE* gene referred to as *E4*. We'll discuss this gene further in Chapter 9, but what's important for our discussion now is the increased risk of Alzheimer's disease after a significant head injury. That risk is increased by about 200 percent in people who have a single copy of the *E4* variant and is increased by about 1,000 percent in people who have two copies.

Genetic screening can identify children who will have an increased risk of Alzheimer's disease if they sustain a significant head injury during their life.

The proper functioning of the *APOE* gene is necessary for maintaining and repairing the structural integrity of the neurons in the brain. When the gene malfunctions (as it does with this variant), that structural integrity is jeopardized, and the brain isn't able to heal itself as efficiently after injury, which means that the brain of a child or young adult is much less resilient to trauma, and Alzheimer's disease is more likely to occur later in life.

A single copy of this variant occurs in approximately 20 percent of African Americans, 15 percent of Caucasians, and 8 percent of Asians. Two copies of the variant occur in approximately 4 percent of African Americans, 3 percent of Caucasians, and less than 1 percent of Asians.

While an everyday knock on the head won't be enough to increase a person's risk of Alzheimer's if he or she has this variant, repeated head trauma from sports or from military service can be enough to lead to dementia or Alzheimer's disease.

THE 88 PLAN

THE ASSOCIATION between head trauma in young adults who have a variant in their *APOE* gene and a substantially increased risk of dementia and Alzheimer's disease is well established. Studies have yet to be conducted to support an association between avoiding head trauma during childhood and reducing the risk of Alzheimer's disease, but I have extrapolated the usefulness of this intervention for children from research dealing with young adults. Besides scientific research, there is also anecdotal evidence of the increased risk in athletes who participate in contact sports, such as professional football.

In 2007, the National Football League started a program called the "88 Plan," which provides up to $88,000 per year to retired players who are suffering from dementia or Alzheimer's and who require financial assistance to pay for their care. The plan is named after the Baltimore Colts Hall of Famer John Mackey, whose jersey number was 88. Mackey, one of the greatest tight ends of all time, endured many head-on collisions throughout his career, and was diagnosed with dementia when he was in his 60s. Many other players have also been affected by dementia or Alzheimer's, some in only their 30s and 40s. A 2007 *New York Times* article on the effects of head trauma in football players stated that "a neuropathologist who examined the brain of Andre Waters, the former Philadelphia Eagles player who committed suicide last fall at 44, said that repeated concussions had led to Mr. Waters's brain tissue resembling that of an 80-year-old with Alzheimer's disease. And last month, the doctors of the former New England Patriots linebacker Ted Johnson, 34, said he was exhibiting the depression and memory lapses associated with oncoming Alzheimer's."

It is because of a rising concern about dementia and Alzheimer's among retired football players that the NFL started the 88 Plan, and Gene Upshaw, the executive director of the players' union, stated that he "was taken aback" when more than 30 retired players signed up for financial assistance immediately after the program was launched.

Sports associations are not only stepping up to help families but are involved in research as well. For example, the Rugby Association is sponsoring grants to study the relationship between head injury and dementia. And preliminary results from an NFL-sponsored study of retired football players has already found that the rate of Alzheimer's disease and dementia is much higher in men between the ages of 30 and 49 who have played professional football compared to those who haven't.

While we don't know whether John Mackey, Ted Johnson, and the other NFL players have the particular variant in the *APOE* gene that would increase their risk of Alzheimer's after head trauma, we do know that not all players are affected. Genetics is the most likely cause, and the *APOE* gene is the first place current research will examine.

The growing awareness of the link among the *APOE* variant, head injury, and dementia is extremely useful information. More than 1 million teenagers play high school football each year, and in 2007, the *New York Times* reported that close to 50 percent of these athletes stated that they had suffered one concussion and 35 percent reported having sustained multiple concussions *in a single season*. This means that close to 500,000 teenagers may be sustaining concussions each year from football. Because the *APOE* variant has an average prevalence of about 15 percent, this equates to 75,000 children with the variant suffering concussions *every year*. And this doesn't include preteens and teens who participate in soccer, ice hockey, wrestling, lacrosse, boxing, snowboarding, skiing, and any other activity that has a high risk of head trauma.

Alerting parents and children to the risk of a disease they might face in the future because of something they do when they are younger empowers them to make informed decisions today.

Children with this *APOE* variant don't have to live in a bubble, but their parents might choose to direct them toward the pursuit of non-contact sports. And both parents and child should be aware of the importance of always wearing a helmet when bike riding, snow-boarding, and so on.

It is interesting that another way to reduce the risk of Alzheimer's is one that will be dear to the heart of any parent—that is, through education. It has been shown that the more one uses one's brain, and the more education one has, the lower one's risk of Alzheimer's. Knowing this, parents can encourage their child to stay in school and study more. And because the protective benefits increase with increasing levels of education, young people predisposed to Alzheimer's might be more inclined to pursue some form of graduate study after college.

GENETIC CAUSES OF ASTHMA

Asthma is a chronic disease that causes inflammation and constriction of a person's airway, leading to shortness of breath and difficulty breathing. It is extremely prevalent, affecting more than 150 million people worldwide and 22 million in the United States, including almost 1 in 10 children under the age of 18. For children living in urban environments, that number rises to almost 1 in 4.

The cause is both genetic and nongenetic, meaning that children may have a genetic predisposition to asthma but that the illness manifests and worsens as a result of exposure to certain nongenetic factors, such as cigarette smoke, traffic-related air pollution, taking certain medications, and even exercise.

Many genetic variants that increase a child's risk of asthma have already been identified, and researchers are now in the process of determining the ways in which potential nongenetic factors interact with these variants to cause asthma. What this means is that if a child has a specific variant, parents can take the steps necessary to limit exposure to the exacerbating nongenetic factor and potentially control or outright prevent the disease. Controlling symptoms and preventing asthma attacks are of paramount importance because uncontrolled asthma can eventually lead to irreversible damage to the child's airway.

Allergens are among the most common nongenetic factors known to be associated with exacerbating asthma. Two of the most common sources of these allergens are dust mites and cockroaches. Cockroaches are quite prevalent, especially in cities, and dust mites, which look like microscopic ticks, live in and consume the dust in our houses. A number of preliminary studies have found associations between specific genetic variants, asthma exacerbations, and exposure to either dust mites or cockroaches.

In 2008, researchers from Harvard Medical School and Brigham and Women's Hospital in Boston published two preliminary studies of children with asthma. One identified variants in the *IL12A* gene as being associated with an increased allergic response to cockroaches. The other found that children with variants in the *IL10* gene had a significantly increased risk of asthma exacerbations when exposed to dust mites while those with asthma but without these variants were much more tolerant of dust mites. The results also indicated that when the concentration of dust mites fell below a certain level, the children with the variants no longer experienced asthma exacerbations.

Both the *IL12A* and *IL10* genes play a pivotal role in the body's immune response to allergens, and variants within these genes cause them to malfunction so that people react more severely to specific allergens. With the use of genetic screening, diseases such as asthma are going to have new, personalized designations. For example, instead of a child being diagnosed with asthma, he or she might be diagnosed with dust mite–intolerant asthma. As a result, preventive measures will become fully personalized and targeted to the exact instigating cause.

Although these variants are associated with environmental factors that contribute to worsening asthma, there are many other variants that are directly associated with increasing the risk of childhood asthma. One of these is in the *ADAM33* gene, which is important for the proper development of the lungs and airway in a growing fetus. Variants in this gene have been shown to increase the risk of asthma, and a recent study found that if a child with one or more variants in their *ADAM33* gene was exposed to cigarette smoke in utero (such as if the mother smoked or was around secondhand smoke while she was pregnant), the child would have a significantly increased risk of asthma.

There are a large number of medications that control asthma, and their effectiveness varies from one person to another. Genetics can help us understand exactly why this happens so that the most effective medications for each person can be prescribed.

Predictive medicine can provide useful information about what medications will be most and least effective in treating asthma.

Variants in the *CRHR1* gene, for example, determine whether inhaled steroid medications (such as Flovent or Pulmicort) will elicit a considerable beneficial response. While all people experience some degree of improvement from these medications, those who have variants in the *CRHR1* gene experience almost four times the improvement in lung function compared to people who don't have any variants.

The *CRHR1* gene is involved in the body's response to stress. When the body experiences stress, *CRHR1* causes the release of cortisol, a protective steroid produced by our body. However, when the *CRHR1* gene contains specific genetic variants, it malfunctions, and the proper amount of cortisol is not secreted in response to stress. Because inhaled steroids help correct this by supplying the body with an external source of protective steroids, people with these variants experience a significant response to the medication. People who don't have the variants, on the other hand, are able to secrete enough of the protective steroid on their own, and, therefore, the addition of external steroids doesn't benefit them as much. The ultimate goal of predictive medicine is to be able to provide people with just this kind of personalized information so that their medical care is tailored to their specific genetic makeup.

THE "TURN YOUR MUSIC DOWN OR YOU'LL GO DEAF" GENES

WE'VE ALL either heard it or said it ourselves: "Turn that music down or you'll go deaf!" Although the risk of going deaf is a bit of an exaggeration (at least most of the time), permanent hearing impairment is actually a valid concern for people exposed to loud music or noise, especially those who have variants in several genes associated with hearing. Usually the noise is work related (if, for example, the person works on a construction site or in a factory that uses loud machinery), but the risk may also be related to repeatedly listening to very loud music, as most children have a propensity to do. For example, noise over 85 decibels is potentially harmful, and the sound on handheld devices such as MP3 players, depending on the headphones used, can get as high as 130 decibels. Because so many people listen to these devices on a loud setting, Dr. Brian Fligor, director of diagnostic audiology at Children's Hospital in Boston and a professor at Harvard Medical School, has stated that there is an unquestionable connection between MP3 players and hearing impairment.

The *HSP70* designation represents a group of genes that protect the body against stress. When an instigating factor, such as acoustic overstimulation from loud noise, occurs, the *HSP70* genes kick into gear and start producing their protective proteins. Variants that decrease the activity of these genes decrease the level of the protective proteins and therefore increase the risk of hearing impairment due to loud noise; those that increase the genes' activity actually protect against hearing impairment under these same conditions.

Genetic screening for a predisposition to hearing impairment as a result of exposure to repetitive, loud noise will alert children and parents to the fact that listening to loud music is likely to have permanent effects on their hearing.

BATTLING CANCER FROM DAY ONE

I strongly believe that prevention of cancer can and should begin in childhood. Instead of waiting decades before taking action, parents and children should be given the opportunity to learn about all preventable diseases for which they are at risk so that they can begin to take preventive measures when the children are still young, even if the disease won't affect them until much later in life. While I talk about the ways predictive medicine can protect adults against cancer in Chapter 10, below I'll discuss how it can also be used to protect our children.

■ BREAST CANCER

You may be surprised to learn that the prevention of breast cancer can begin in childhood. This is primarily accomplished by exerting control over nongenetic factors such as exposure to radiation from x-rays and CAT scans. We've known for some time that exposure from radiologic tests will increase a woman's risk of breast cancer for *decades* after the test is performed, but additional studies are now showing that women who are genetically predisposed to breast cancer, such as those with variants in their *BRCA1*, *BRCA2*, *CHEK2*, or *ATM* genes, will be at even *higher* risk after radiation exposure. In fact, studies have found that if these women have even low-dose radiation, such as that from chest x-rays, before the age of 20, their risk will be further increased by more than 250 percent.

Many of the variants that increase a person's risk of breast cancer do so because they decrease the effectiveness of genes that are responsible for repairing DNA when it becomes damaged. And radiation is a significant cause of DNA damage. The more DNA damage these people incur, the greater the likelihood that cells in the breast will start to divide uncontrollably, eventually leading to cancer. So, while these variants increase the risk of breast cancer even without radiation exposure, radiation causes the risk to skyrocket by throwing fuel on a fire. Now that we have this information and the ability to screen for all of these variants at once, it is, in my medical opinion, extremely prudent to limit the radiation exposure of children (girls *and* boys) who have one or more of these breast cancer variants from the moment they are born. Much of the information that is obtained from tests that expose a person to radiation can be ascertained with other tests, such as more extensive physical exams, blood tests, MRIs, and ultrasounds, that

Children who are genetically predisposed to breast cancer should avoid exposure to radiation.

involve absolutely no radiation at all. If, for example, a doctor thinks your child might have a lung infection, instead of automatically sending him or her for a chest x-ray, the physician could base the diagnosis on a physical exam and blood tests.

Of course, in cases of a life-threatening emergency there may be no way to avoid an x-ray or CAT scan, but if the radiologic test is elective, knowing whether your child has a breast cancer genetic variant can influence the choices a doctor makes about how best to provide medical care for that child.

■ SKIN CANCER

More than 1 million cases of skin cancer are diagnosed in the United States each year. Skin cancer occurs in two forms, melanoma and nonmelanoma, the latter of which includes basal cell and squamous cell skin cancer. Although the majority of deaths related to skin cancer result from melanoma, all skin cancer must be removed surgically, often causing pain and, in some instances, disfigurement.

Almost all skin cancers develop due to a combination of genetic and nongenetic factors, and the primary nongenetic factor is almost always too much sun exposure. Each sunburn, especially before the age of 20, significantly increases one's risk of developing skin cancer. Because of this, avoiding sunburns during childhood is extremely important for reducing one's overall risk. And just as with breast cancer, people who have certain genetic variants are at an even higher risk than they would be otherwise if they are exposed to too much sun during childhood.

Clearly, then, if parents know that their child has one of these skin cancer variants, they can take extra precautions to limit the child's sun exposure throughout childhood and make sure to aggressively protect him from sunburn. We've all been told that we ought to wear sunscreen, but for some people that advice can be a lot more meaningful than it is for others. Predictive medicine can provide insight into whether a parent needs to be extra-vigilant in protecting a child from the sun.

Predictive medicine can provide parents with insight about their child's risk of skin cancer.

Conveying an increased risk of skin cancer is also important for adolescents who might otherwise be using tanning beds, which increase the risk for skin cancer just as much as direct exposure to the sun. Close to 40 percent of teenage girls and 12 percent of teenage boys use tanning booths, even though they are now classified as a Group One cancer-causing agent by the International Agency for Research on Cancer, meaning they are equivalent to arsenic, asbestos, and smoking. Informing your child of his or her genetic risk for skin cancer may be just the motivation the child needs to consistently avoid indoor (and outdoor) tanning.

▪ LUNG CANCER

Lung cancer is not only the most prevalent but also the deadliest of all cancers, with close to 1.5 million new cases and just over 1 million deaths recorded throughout the world every year. Lung cancer is directly related to smoking and secondhand smoke, but it also has a significant genetic component, and a plethora of genetic variants have been associated with a person's increased risk of the disease. In this case, however, genetics determines only about 8 percent of the risk, so environmental factors greatly outweigh genetics. As we all know, however, cigarette smoking due to nicotine addiction is the primary nongenetic factor associated with lung cancer, and about 60 percent of whether a person will become addicted to cigarettes *is* determined by our genes, with 40 percent being determined by nongenetic factors, such as whether or not the person's parents are smokers. Therefore, genetics does ultimately play a substantial role in determining one's risk for lung cancer.

It would be almost impossible for any child or adult not to have heard that smoking is bad for one's health. And yet 50 percent of all high school students have smoked cigarettes in the past and about 20 percent are currently smokers. A 2006 survey conducted by the CDC also found that 1 in 10 *middle school* children had used tobacco products within the last 30 days. And close to 90 percent of all adult smokers initially started smoking when they were teenagers. Research reveals that if a person doesn't start smoking as a child or a teenager, he or she is unlikely ever to start. Based on this information, it should be clear that preventive measures aimed at stopping lung cancer and cigarette smoking need to focus on children.

Now, instead of simply telling children that smoking is bad for them, we can say, "Based on *your* genetic code, *you* are at increased risk for becoming addicted to nicotine, so if you start smoking, you may never be able to stop, and smoking may lead to your developing lung cancer."

Identifying children who are at risk of nicotine addiction allows for preventive measures to be instituted *before* the child starts smoking.

The salient point here is that early intervention based on genetic information is what's going to have the greatest impact on your child's long-term risk of lung cancer. If we can successfully stop some children who have a predisposition to nicotine addiction from smoking, we'll already have made significant headway in our battle against lung cancer.

Examples of Child-Related Genetic Screening Panels

Newborn Panel

- Personal genetic identifier
- Blood group
- Effect of breastfeeding on intelligence (IQ)
- Lactose intolerance
- Heart abnormalities (including arrhythmias and other preventable causes of sudden death including SIDS)
- Full pharmacogenomics profile (including dosing, effectiveness, and adverse reactions to medications that the newborn may potentially be given throughout childhood)
- Multiple sclerosis risk with low vitamin D levels
- Alzheimer's disease risk with head injury
- Cancer risk with radiation, sun exposure, secondhand smoke, and certain foods and beverages
- Pervasive developmental disorder (including autism, autism spectrum disorders, Asperger's syndrome, and Rett's syndrome)
- Blood clotting disorders
- Pyloric stenosis (abnormality of the stomach)
- Neonatal diabetes
- Rare disease screen (including orphan diseases, metabolic diseases, and syndromes)

Children's Panel

- Personal genetic identifier
- Blood group
- Heart abnormalities (including arrhythmias and other preventable causes of sudden death)
- Dyslexia
- Reading performance
- Pervasive developmental disorder (including autism, autism spectrum disorders, Asperger's syndrome, and Rett's syndrome)

- Obesity and leanness
- Athletic ability and predisposition to specific sports
- Genetic age and effectiveness of exercise regimens
- Sickle-cell trait (carrier of sickle-cell disease)
- Multiple sclerosis risk with low vitamin D levels
- Breast cancer
- Ovarian cancer
- Skin cancer (both melanoma and nonmelanoma)
- Nicotine addiction
- Brain aneurysm
- Asthma
- Allergies
- Full pharmacogenomics profile (dosing, effectiveness, and adverse reactions to medications that the child may potentially be given throughout childhood)
- Anesthesia requirements for proper sedation
- Lactose intolerance
- Noise-induced hearing impairment
- Susceptibility to and severity of infectious diseases (including meningitis, traveler's diarrhea, Lyme disease, West Nile virus, HIV, tuberculosis, and stomach flu)
- Taste perception and food preferences
- Blood clotting disorders
- Visual acuity (including color blindness, night blindness, Leber congenital amaurosis, and macular degeneration)
- Chronotype (whether the child is predisposed to perform better during the daytime or at night)
- Narcolepsy
- Effect of stimulants on cognition
- Effect of caffeine consumption on sleep

For additional panels related to newborns, children, and teenagers, please visit www.OutsmartYourGenes.com/Panels.

8

Protecting Your Cardiovascular Health by Beating Your DNA

MISCONCEPTION: Cardiovascular health is all about eating right and exercising. As long as you do both on a regular basis, your heart and blood vessels will be healthy.

FACT: Sometimes, no matter how good you are to your body, your body may not be good to you. Many people have genetic variants that increase their risk of cardiovascular disease no matter what their lifestyle. While there are many preventive measures available, decreasing your risk of cardiovascular disease will be much more effective if you know which variants you have *before* you put protective measures in place.

PROTECTING THE BEAT OF LIFE

Heart disease, closely followed by cancer, is still the number one killer worldwide. Because of this, predicting and preventing cardiovascular disease is a cornerstone of predictive medicine. In the United States, 40 percent of people between the ages of 40 and 59 years, 70 percent of people between 60 and 79 years, and approximately 80 percent of people who are 80 and older have at least one cardiovascular disease.

In fact, it's likely that one person has already died from cardiovascular disease in the time it's taken you to read this page. In the United States, one person dies every 37 seconds from a disease affecting the heart or the circulatory system, and approximately *1 in every 3* people worldwide dies of cardiovascular disease each year (that's around 17.5 million people annually). Even though life expectancy in the United States is almost 80 years, 30 percent of the deaths attributed to cardiovascular disease occur *before* the age of 75, which means that a lot of people are dying prematurely because of problems with their heart or blood vessels. The economic consequences of these statistics are colossal, with the National Bureau of Economic Research stating that the total cost of cardiovascular disease in the United States alone is now more than $110 billion *each year*.

A tremendous amount of research and money has been spent to develop medications and procedures to treat almost any kind of heart-related problem, but to treat a problem successfully we have to know it exists—before it's too late. Predictive medicine harnesses the power of genetic screening to identify people at risk of cardiovascular disease years to decades before any symptoms, or even any disease, manifest so that genetically tailored preventive measures and treatments can be instituted at the point when they will be most effective.

THE GENETICS OF LONGEVITY

In the field of genetics, the term *longevity* refers to life expectancy under ideal conditions. Longevity runs in families—children of long-lived parents are more likely to live longer themselves, while in other families it's known that, unfortunately, people just seem to die young.

When I first started to integrate genetics into my medical practice, I thought that the genetics of longevity must surely encompass a large number of genes responsible for a variety of bodily functions. But it soon became apparent that longevity genes didn't come in all that many flavors. In fact, it is primarily the genes affecting the cardiovascular system that determine our longevity. If variants that affect the heart are beneficial, you live longer; if they're harmful, you die prematurely. This isn't true of variants that affect any other organ in your body; only those that affect your heart or blood vessels are directly associated with longevity.

THE FRENCH PARADOX
Alcohol, Red Wine, and Resveratrol

THE FRENCH paradox refers to the fact that even though the French diet contains a significant amount of saturated fat, the French people don't suffer the high rate of heart disease that would normally occur with that type of diet. In fact, the rate of heart disease in France is *50 percent lower* than that of the United States and the United Kingdom even though all of their risk factors (including blood pressure, cholesterol levels, smoking, and obesity) are the same. The answer to the paradox lies in the abundance of charming cafés that line the Avenue des Champs-Élysées in Paris: The French diet also contains a daily dose or two of red wine.

It's a widely accepted medical fact that moderate consumption of beer and red wine protects against heart disease. The key word, however, is *moderate*, because the benefits from alcohol consumption follow a J-shaped curve: The higher up you go on either side, the greater your risk of death, and the lower you are, the lower your risk of death. If you are able to limit alcohol intake to one to two glasses maximum per day, your health may benefit, but if you drink any more than that, it would be better for your health to drink no alcohol at all.

On the left side of the J are people who don't drink at all, and their risk of death from heart disease is about 40 percent higher than that of people at the lowest part of the J, who are those who drink one or two alcoholic beverages a day. On the right side of the J are people who drink more than two alcoholic beverages a day; their risk of death from cardiovascular disease shoots up significantly, to much higher than that of people who don't drink at all. So, the goal would be to drink moderately— about 5 ounces of wine or 12 ounces of beer per day for women and double that for men.

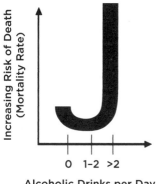

Alcoholic Drinks per Day

The term *French paradox* was first proposed by Serge Renaud, a medical researcher at France's National Institute of Health and Medical Research (INSERM), who, appropriately enough, grew up on a vineyard near Bordeaux. While there are certainly health benefits with moderate consumption of beer and other types of alcohol, the French paradox relates primarily to the benefits of red wine, which is different from other types of alcohol because of substances called polyphenols, which are contained in the skins of the grapes. (Note that when making white wine the skins are removed before fermentation begins, while in red wine they are left in throughout the entire process.) Polyphenols have a number of health benefits: They thin the blood by making platelets less sticky, release chemicals that allow the heart to pump more efficiently, contain antioxidants that help take the badness out of bad cholesterol, decrease inflammation in the blood vessels, and precondition and protect the heart and brain so that they are more resistant to injury if it should ever occur. Wine also contains a number of other active substances, including flavonoids, which are also found in lesser concentrations in many fruits, vegetables, and teas, that further increase its health benefits.

continued

Because wine contains these beneficial substances, it's been shown not only to decrease the risk of heart disease by about 40 percent but also to decrease the risk of both lung and prostate cancer by 50 percent. And it may also reduce the risk of melanoma and other types of cancer.

The polyphenols and flavonoids are what separate red wine from other alcohols in terms of its range of health benefits. That's not to say that beer consumption doesn't have health benefits, only that red wine consumption has more.

Although some white wines are also infused with polyphenols, the majority are not, but the white wines that are polyphenol infused appear to have similar health benefits as red wine. But because most white wines lack polyphenols, the health benefits of reds appear to outweigh those of whites.

One of the polyphenols whose protective qualities have received the most attention is resveratrol, a word you may already have heard because it's received a bit of media attention as a potential anti-aging supplement. Red wine is one of the richest natural sources of resveratrol, which is known to make the heart and brain more resilient if injury occurs.

How does this relate to your genes? Recent research has found another way resveratrol may be exerting its beneficial effects on the body, and that is by causing the SIRT1 gene to become more than 1,000 percent more active than it would be normally. In essence, it supercharges the SIRT1 gene, which is directly involved in anti-aging. The SIRT1 gene produces an enzyme that makes cells more stress resistant both by augmenting the cell's ability to heal itself and by delaying it from self-destructing so that it has more time to repair itself and live on instead of dying.

In studies of animals, resveratrol has been found to extend longevity by up to 50 percent, while preserving the animals' physical and mental capabilities. It appears to protect animals from both Alzheimer's disease and Lou Gehrig's disease so that they not only live longer but also stay mentally sharp and physically healthy. And the amount of resveratrol found in red wine has been shown to be enough to cause these effects. Because of this, it now appears that red wine goes straight to the DNA and supercharges our anti-aging genes!

As a result of these findings, there are now a number of companies offering resveratrol, purified from the skin of grapes, in pill form. Before you rush off to the store to buy resveratrol pills, however, you should understand that what has been found in animal studies might not necessarily translate to humans. As many pharmaceutical companies have found in the past, a medication that works in animals does not always work the same way in humans. This doesn't mean that resveratrol *doesn't* work in humans; it just means that rigorous human studies are needed before supplementation can be recommended. Meanwhile, we already know that drinking red wine in moderation does have significant health benefits. And if red wine isn't your thing, red grape juice has a high concentration of resveratrol too.

OUTSMARTING A FRAILTY GENE

The *APOE* gene is one of those most studied in relation to longevity. The *APOE* gene plays a role in how your body processes cholesterol, and variants within this gene are associated with an increased risk of premature death. One study found one of these harmful variants in about 20 percent of their study population between the ages of 18 and 63 but in only 11 percent of people between the ages of 64 and 109, indicating that those who have the harmful variant are at a much greater risk of death once they reach the age of 60.

Because of this, some geneticists prefer to call *APOE* a "frailty gene" rather than a longevity gene. The frailty associated with decreased life expectancy occurs because the variant interrupts the gene's proper functioning so that cholesterol is not broken down appropriately by the person's body. People with this variant are, therefore, at increased risk of having cholesterol clog their arteries, a condition called coronary artery disease (CAD). Because the arteries feed blood to the heart muscle, if a blockage occurs, blood can't feed the heart, and a heart attack ensues.

The increased risk of death due to harmful *APOE* variants begins to manifest at about age 40 (when you're at slightly increased risk) and then quickly escalates until, after the age of 60, your risk of death increases by more than 300 percent compared to that of the general population. In terms of predictive medicine, this is tremendously useful information because we know what will happen (clogged arteries) if a person has one of these variants, and we know approximately when the damage will start to occur (after the age of 40).

Knowing this, we can start to institute preventive measures between the ages of 30 and 35 by performing annual blood tests to check both levels of cholesterol and indicators of inflammation. We can also start patients on a cholesterol-lowering medication, and we can conduct screening tests to make sure the arteries around the heart stay clog free. Just knowing that you are at increased risk for

There are many ways to lower your risk if you are found through genetic screening to be predisposed to heart disease.

heart disease may be the kick in the butt you need to see your doctor on a regular basis and for your doctor to make sure to call you if he or she hasn't seen you in more than a year.

Some physicians may say that starting a cholesterol-lowering medication called a statin (such as Lipitor, Zocor, and Crestor) even when cholesterol levels are normal is not warranted or indicated. But some excellent studies have already shown that preventive treatment of pre-disease states can have tremendous benefits. For example, one study conducted by the director of the Center for

Cardiovascular Disease Prevention at Harvard Medical School's Brigham and Women's Hospital investigated the effectiveness of prescribing the cholesterol-lowering statin medication rosuvastatin (Crestor) to prevent heart attacks, strokes, and overall death in close to 18,000 people. The study found that people with *normal* cholesterol levels and no other risk factors for cardiovascular disease but with elevated levels of CRP (a biomarker indicating that there is inflammation occurring somewhere in the body) had *significantly* reduced risks of disease and death when started on a statin. Those who took the drug had a 54 percent lower risk of heart attack, a 48 percent lower risk of stroke, a 53 percent lower rate of hospitalization, a 47 percent lower risk of death from cardiovascular disease, and a 20 percent lower risk of death from *any* cause whatsoever.

Starting cholesterol-lowering medications even if you have normal cholesterol levels may significantly reduce your risk of heart disease and death.

The name of this study was the Justification for the Use of Statins in Prevention: An Intervention Trial Evaluating Rosuvastatin (JUPITER). In fact, the study did just that—it justified *prevention* of cardiovascular disease and death through pre-disease intervention by proactively prescribing a medication, even to low-risk people, instead of waiting for overt signs of a disease to manifest.

Another study investigating the direct action of prescribing a statin (Zocor) to people who had a harmful *APOE* variant found that the statin successfully reduced the risk of death by more than 70 percent. This study exemplifies the power of predictive medicine by showing that a variant known to be associated with increased risk of death can be outsmarted simply by prescribing a common medication.

PREDICTIVE SYNERGY
Genetic Screening and Biomarkers

A BIOMARKER is a substance in the body that can be used to detect disease days, weeks, or even months before the first symptoms appear. As discussed earlier, elevated levels of the biomarker CRP indicate inflammation somewhere in the body, which is associated with an increased risk of coronary artery disease and heart attacks. Most diseases exist for a significant amount of time before we become aware of them, and during this time underlying changes related to the onset of the disease are usually occurring in our body. A tremendous amount of research has been aimed at understanding these biomarkers so that they can be used to predict the presence of underlying disease and, therefore, allow us to implement treatment before the disease has time to take hold.

This may seem very similar to genetic screening, and while it is an extremely important part of personalized medicine, there are a few key differences. Most notably, biomarkers indicate the presence of disease, but they do not predict or detect disease *before* the disease process occurs.

Another important difference is that our DNA is the fundamental code of life, but biomarkers are not. Many biomarkers are proteins that are *produced by* genes, so biomarkers, just like everything else in our body, vary according to our genetic makeup.

There are, for example, variants within the *CRP* gene that alter its ability to produce its protein, causing the person to have either higher or lower than normal levels of CRP. Therefore, if biomarker levels are analyzed without considering genetics, false assumptions may be made. If you have variants that cause lower CRP levels and you also have coronary artery disease, your CRP levels could be as much as 30 percent lower than they usually are in people with the disease, indicating that you don't have coronary artery disease even when you do.

Alternatively, there are other variants that cause CRP levels to be higher than normal, so even if you don't have an underlying disease, test results for the CRP biomarker could falsely indicate that you do. However, once you are aware that your DNA contains variants in the *CRP* gene, the cutoff values for determining whether you have a disease can be adjusted to create what is called a genetically calibrated biomarker. This applies not only to CRP but to a wide range of biomarkers.

I believe that predictive medicine will be able to use genetics in conjunction with genetically calibrated biomarkers to significantly increase our ability to predict disease risk and then monitor for that disease so that if it ever manifests we will detect it and intervene as early as possible.

USING PREDICTIVE MEDICINE TO AVOID SUDDEN DEATH

The heart uses a flow of electrical impulses to coordinate its beat. As with any electrical system, it has a generator (the natural pacemaker) from which the impulse originates, rapidly circumventing the heart and telling the muscle to contract. When it contracts, the heart pumps blood throughout the body. The heart's electrical impulse, therefore, is responsible for the beat of life. Simply put, without it we die.

How our electrical conduction system works is dictated by our genes. And variants in particular genes can cause a person to be predisposed to momentary lapses in his or her electrical conduction. If the electrical rhythm is off even by a millisecond, this can lead to an irregular rhythm of the beat that can then rapidly deteriorate into uncoordinated beating that doesn't pump blood efficiently. When this occurs, the person has only moments to live. Therefore, variants in the genes that are associated with our heart's electrical system put a person at risk for sudden death. Learning that a person carries one of these variants can be life saving because there are a number of interventions that can be used to prevent the electrical system from misfiring. Without that foreknowledge, however, there are no second chances. The only way to prevent sudden death is to intervene before it occurs.

Both medical journals and the popular media abound with accounts of children and adults dying suddenly and unexpectedly because of an abnormality that affected their heart but was not detected until it was too late. Among the many cases described in the literature are those of a 9-year-old girl who died suddenly while swimming, a 14-year-old girl who died while she was doing nothing more than talking on the phone, a 17-year-old boy who died while playing soccer, and a man who died in his sleep at the age of 50. In almost every instance, these victims of sudden death were considered healthy and disease free. Their deaths seemed to come out of the blue. But in actuality they were all carrying undetected variants in their genetic code that predisposed them to sudden death.

Abnormalities in the rhythm of the heart beat (known as arrhythmias) are primarily responsible for the more than 450,000 sudden deaths each year in the United States alone. Because physical exertion is a major trigger for a life-threatening abnormal heart beat in people predisposed to arrhythmias, athletes are especially at risk. Every so often you'll hear the unfortunate story of an athlete who suddenly collapsed and died when playing a sport or exercising, and about 40 percent of the time these deaths occur because the person had an undetected but insidious condition called hypertrophic cardiomyopathy (HCM). There are more than 10 genes associated with HCM, and all it takes is a single variant in just one of them to cause the wall of the heart to thicken, which can then cause an instability in the conduction of the electrical impulses, especially when the heart is beating fast during physical activity, leading to an arrhythmia and sudden death. About 1 in 500 people have HCM, which means it occurs too frequently to be considered a rare disease.

Another disorder that often goes undetected but that can cause sudden death, especially in athletes, is called arrhythmogenic right ventricular dysplasia (ARVD). There are more than seven genes associated with ARVD, and having just a single variant in any one of them may cause the disease. The variants disrupt the normal scaffolding that connects one heart cell to another, which can affect the proper conduction of electrical impulses throughout the heart. In the United States, ARVD is responsible for about 20 percent of all cases of sudden death in people under the age of 35, and about 60 percent of these people have no family history of the disease.

In Italy, which currently screens athletes for disorders that can cause sudden death, the rate of sudden death among athletes has been reduced by an astonishing *90 percent*. Because anyone who has one of these arrhythmia-causing variants is at risk of sudden death regardless of whether he or she is an athlete, predictive medicine is able to screen for all genetic causes of sudden death in *both* athletes and nonathletes.

In addition to being triggered by physical exercise, other arrhythmias related to particular genetic variants can be tied to other instigating factors, including common medications. Without these triggers, no arrhythmia is likely, but when a person predisposed to an arrhythmia comes into contact with one of these factors, the chances of experiencing an arrhythmia skyrocket.

As an example, researchers at Harvard Medical School performed an in-depth study of the *SCN5A* gene and its association with arrhythmias. The *SCN5A* gene is involved in creating the electrical signal that pulses throughout the heart, telling it to beat. Researchers have already

Predictive medicine is able to identify genetic variants that can cause fatal arrhythmias of the heart and lead to sudden death.

identified more than 100 different variants that cause the *SCN5A* gene to malfunction, and although most of them are rare, one in particular occurs in approximately 1 in 8 African Americans and 1 in 5 blacks from West Africa and the Caribbean. It has also been found in Caucasians, Asians, Hispanics, and other populations, but it is by far most common in blacks. The Harvard study found that a single copy of this variant, which may exist in more than 4.5 *million* African Americans, increases a person's risk of an arrhythmia by more than 800 percent, primarily if he or she also comes into contact with or has certain non-genetic triggering factors: particular medications; improper nutrition; kidney or liver disease; sleep apnea; and any event that may cause fluctuations in the body's potassium level, such as vomiting, diarrhea, and surgery. The same variant also appears to substantially increase an infant's risk of dying from SIDS, but, again, primarily when the infant is exposed to instigating factors.

The discovery of this variant is highly significant for two reasons. First because it is consistent with and may partly explain prior research indicating that there is a higher prevalence of arrhythmia-associated sudden death in blacks, and second because being able to detect the variant allows for the enactment of preventive mea-

Many life-threatening arrhythmias occur when a person with a genetic variant comes into contact with specific nongenetic triggering factors.

sures that are likely to avert sudden death. The authors of the study summed it up beautifully when they wrote, "The key to therapy is prevention." Preventive measures would include, among others, regular monitoring by a cardiologist, possibly taking a beta-blocker, and being educated about specific instigating factors to avoid.

PROTECTING YOUR ARTERIES WITH PREDICTIVE MEDICINE

Arteries are the passageways through which oxygen-rich blood travels from your heart to all the organs in your body. Some arteries also circle back to feed the heart muscle itself. If an artery feeding the heart muscle becomes clogged, you'll have a heart attack and part of the heart will die. If a blockage occurs in an artery feeding the brain, you'll have a stroke and a part of the brain will die. High cholesterol levels, which cause blockages, heart attacks, and strokes, are all caused by an interaction between nongenetic factors (what we eat, if we smoke, and how much we exercise) and genetic variants.

More than 40 percent of all cardiovascular disease-related deaths worldwide are caused by heart attacks, and the majority of these are caused by a buildup of cholesterol within our arteries. Hundreds of genetic variants have been associated with high cholesterol levels. As an example, multiple research studies have shown that commonly occurring variants within the 9p21 region of chromosome 9 are associated with coronary artery disease, which is the disease that occurs when cholesterol builds up in the arteries that supply blood to the heart. Approximately one-quarter of all people have two copies of these variants, which increases the risk of CAD and heart attacks by more than 75 percent for people under the age of 55.

What does "within the 9p21 region" mean? This is genetic terminology for a specific region on chromosome 9 and is commonly used when variants are not within a gene. This can occur because genes don't exist back-to-back and there are usually expansive regions *between* one gene and the next. Because these regions don't serve any known function, for a long time they were referred to as "junk DNA." As it turns out, however, the DNA in these regions is far from being junk, as can be seen from the variants in the 9p21 region we're now discussing.

Because CAD and heart attacks cause so many deaths, physicians commonly use standard formulas that take into account multiple risk factors for these diseases and provide a single score that equates to a risk assessment. While numerous risk factors—such as age, body mass index, blood pressure, cholesterol levels, and tobacco use—are taken into account, most genetic variants are not because they've been only recently discovered. However, a study that integrated genetic variants into one of the most commonly used heart disease risk formulas found that, based on information for just a single genetic variant in 9p21, more than 1 in 8 people were instantly reclassified into a higher or lower risk category. This implies that genetic information is not only useful but necessary for properly assessing a person's risk of heart disease.

Integrating genetic information into traditional formulas used to predict risk of heart disease significantly improves their accuracy.

YOUR *DENTIST* MAY BE YOUR HEART'S BEST FRIEND

THE MEDICAL literature is full of interesting associations, such as a peculiar link between periodontitis and heart disease. Periodontitis is an inflammation of the gums that surround and support the teeth. Left untreated it can lead to the destruction of part of the jawbone and loss of teeth. It's actually an autoimmune disease that occurs when a person's immune system becomes too aggressive in attacking the bacteria around the teeth and starts to target the gums as well. In the United States, approximately 1 in 3 people suffer from periodontitis sometime during their life, with 1 in 10 having a severe form.

Numerous studies have shown that people with periodontitis are at greater risk for coronary artery disease, heart attacks, and strokes. It also appears that the severity of the two are linked so that people with more severe periodontitis also have more severe heart disease. And, even more significant, studies have shown that if you properly treat periodontitis, you can actually improve the functioning of the arteries in the heart and brain, decrease cholesterol levels, and reduce your risk of cardiovascular disease.

One hypothesis in regard to this association is that the inflammation associated with periodontitis has far-reaching effects throughout the body, causing platelets to stick together and plaques to form in the blood vessels. In fact, bacteria from people's teeth have actually been found in the plaque lining their coronary arteries.

Our risk of periodontitis, like our risk for heart disease, is determined by both genetic and nongenetic factors. And just recently it was discovered that a single variant in the 9p21 region increases the risk of both heart disease and periodontitis. If a person's genetic makeup contains two copies of this variant, the risk for both diseases increases by more than 75 percent.

Because of their shared origins, it's not surprising that both periodontitis and cardiovascular disease have many of the same modifiable risk factors, including cigarette smoking, diabetes, and obesity. And now it appears that cardiovascular disease has a new modifiable risk factor—dental health. Your risk of periodontitis can be decreased with a conscientious adherence to daily dental hygiene such as brushing at least twice a day with a soft-bristle toothbrush, flossing, and using antiseptic mouthwash. It is also important to have regular checkups with a dentist because protecting yourself against periodontitis will also decrease your risk of cardiovascular disease. Your dentist may just be your heart's new best friend.

There are many variants in a number of genes that increase a person's risk of blood clots inappropriately forming in a blood vessel and when this occurs in a vein in the leg, the blood clot is referred to as a deep vein thrombosis (DVT). DVTs can be deadly because, although they form in a vein in the leg, they may dislodge and travel to the lung, where they can cause sudden death. When lifestyle factors such as smoking, obesity, a sedentary lifestyle, or taking medications that contain estrogen (such as oral contraceptive pills and hormone-replacement therapy), are combined with these variants there is a synergistic effect that increases the risk of harmful blood clots forming *even further*.

As always, if you know you have the variants, you can be empowered to lower your risk of blood clots and heart attacks through lifestyle modifications such as stopping smoking, losing weight, becoming more physically active, and avoiding medications that contain estrogen. And you can take advantage of devices that painlessly push against the outside of your legs to keep your blood circulating and clot free.

These measures could perhaps have helped prevent the death of David Bloom, the young and seemingly healthy NBC correspondent who died suddenly of a blood clot that traveled from his leg to his lungs in 2003 when he was covering the first war in Iraq. It was only after his death that genetic testing revealed he carried one of the variants that increases the risk of blood clots. Although this specific genetic variant usually isn't associated with life-threatening blood clots, it has been known to cause death when other, instigating factors are involved, such as prolonged immobilization. Bloom's death was most likely due to a combination of his genetic predisposition, the long plane flight he had recently taken, and his severely restricted mobility of sitting in the same cramped position with his knees against his chin for long periods of time while riding in a military tank.

If he had been aware of his predisposition, he would not only have been able to take precautions (such as not riding in a cramped tank for long hours) but would also have been alerted to act on a key symptom associated with DVTs—the leg cramps he had been complaining of for a few days before his death.

REDUCING STROKE RISK AND AVOIDING DISABILITY

A stroke occurs when blood flow to the brain is disrupted, thereby depriving some of your brain cells of oxygen and causing them to die. While the vast majority of strokes are caused by a buildup of cholesterol in the arteries, similar to what occurs with a heart attack, some may occur when a blood vessel becomes weakened or the blood itself becomes too thin and actually spills out of the vessel and into the brain.

In the United States, stroke is the third leading cause of death and the number one cause of long-term disability; 75 percent of stroke victims suffer some form of disability if they survive. About 1 out of every 6 people will have a stroke sometime during their life. Strokes also have significant consequences for our economy, costing the United States more than $65 billion each year. With prediction and prevention, you *can* lower your risk of stroke.

Earlier in this chapter we discussed the *APOE* gene in relation to heart attacks, but because it is responsible for how we break down cholesterol, the same variant is also responsible for an increased risk of stroke. And that mysterious 9p21 region, which is associated with CAD, heart attacks, and even periodontitis, is also associated with a significant increase in risk of stroke. We've already discussed arrhythmias in relation to sudden death; some arrhythmias, however, aren't directly life threatening but can cause serious harm by increasing the chances of a stroke. In fact, one of the most significant risk factors for a stroke is an arrhythmia of the heart called atrial fibrillation (usually referred to as a-fib).

A-fib is the most common of all arrhythmias, affecting approximately 2 percent of people over the age of 55 and close to 10 percent of all people older than 80. The disease is characterized by an abnormal rhythm in the beating of the heart that prevents blood from being pumped out as effectively as it should be. As a result, blood pools and sticks around inside the chamber of the heart for much longer than it's supposed to. This pooling can lead to the formation of blood clots, which may then be pumped out of the heart to travel throughout the body. If one of these clots makes its way to the brain and becomes lodged in a blood vessel, it can cause a stroke. Although some people have noticeable symptoms, such as heart palpitations or a rapid heart beat, many times people don't even know they have a-fib. And although it can easily be detected during a routine physical exam if it is occurring at the time, a person can go in and out of a-fib at random intervals, so it may not be happening at the precise moment the examination is taking place.

While a-fib can be caused by high blood pressure and other conditions that affect the heart, one of its significant genetic causes are variants that affect the *PITX2* gene, which is involved in making sure the heart's electrical impulses are conducted properly. It is not surprising that a recent study found that these same variants were also associated with an increased risk of stroke. But one study has found that many of the people who have

A predisposition to a-fib, one of the most common causes of strokes, can be predicted with genetic screening.

one or more of these variants and suffer a stroke had *not* been previously diagnosed as having a-fib. Therefore, identifying people with these variants can alert both the person and the physician to be on the lookout for previously undiagnosed a-fib. Because a person with the condition has a 35 percent risk of having a stroke, and about 20 percent of all strokes are caused by a-fib, diagnosing and treating the arrhythmia as promptly as possible can considerably decrease the chance of a stroke. Treatments that reduce a person's risk of stroke include medications and interventions to stop the a-fib or to prevent the formation of blood clots.

SEXIST GENES AND . . . TRIGLYCERIDES

BELIEVE IT or not, some genetic variants discriminate on the basis of sex. In the field of genetics we refer to these variants more politely as sex dependent.

In 2007, researchers at the University of California Los Angeles (UCLA) School of Medicine published a study that brought to light sexist genetic variants within the *USF1* gene, which is involved with lipid and triglyceride metabolism. Triglycerides are a type of fat, similar to cholesterol. High levels of triglycerides, like high cholesterol levels, are linked with clogging of the arteries. What the study found is that a specific variant in this gene is associated with high triglyceride levels only in men, while *not* having the variant was associated with high triglyceride levels only in women. This may seem odd, but this finding has actually been corroborated by a number of other studies.

While other medical research has noted clear differences in the occurrence of coronary artery disease and heart disease in men and in women, this study was one of the first to observe such a sex-dependent difference attributable to a single genetic variant. The reason for this variant's sex-dependent nature is still unknown, but one hypothesis relates it to the differences in hormone levels between men and women.

AVOIDING BRAIN ANEURYSMS

A brain aneurysm is a weakening in the wall of a blood vessel that runs through the brain, causing the blood vessel either to enlarge abnormally or to balloon out because of the pressure of the blood flowing through the vessel. Brain aneurysms, which affect 1 out of 50 people, are extremely dangerous because they are likely to rupture, and when they do, the person suffers a devastating stroke, often associated with severe disability or death. Factors that significantly increase the risk of a brain aneurysm are high blood pressure, cholesterol buildup in the arteries feeding the brain, head trauma, and, genetics.

In 2008, researchers at Yale University published a study that identified many of the genetic variants associated with an increased risk of brain aneurysms. They found that a number of variants found at three different locations in the genetic makeup of a person increase the risk of stroke substantially and account for approximately 40 percent of an individual's total genetic risk for brain aneurysms. Outsmarting these variants will significantly decrease stroke, disability, and death from brain aneurysms.

The variants, including a few in the 9p21 region, most likely cause a malfunctioning of genes that appear to be involved not only in the proper construction of the walls of the blood vessels but also in the walls' maintenance and repair. When these genes aren't working properly, blood vessels are weak and unable to repair themselves if damage should occur. When an instigating factor, such as years of high blood pressure, smoking cigarettes, or head trauma, occurs, the already weakened blood vessel walls are exposed to greater stress, and this can cause the formation of an aneurysm.

A number of preventive measures can be instituted to decrease the risk of a brain aneurysm as well as the risk of serious disability and death should an aneurysm ever manifest.

Fortunately, there are many preventive measures that can be instituted to decrease the risk of a brain aneurysm and allow for early intervention if one should occur. For example, knowing that a person is at increased risk for a brain aneurysm will alert the person's physician to be very aggressive in monitoring and controlling blood pressure and, if the person is a smoker, instituting therapies to help the person quit. In addition, if parents know their child has one of these variants they might guide him or her to avoid contact sports and wear a protective helmet when bike riding or Rollerblading. The point, as always, is to give people the insight about their genetic makeup that will allow them to make informed decisions years or decades *before* the disease strikes.

AVERTING HEART FAILURE

Heart failure occurs when the heart can no longer pump blood effectively throughout the body. Most of the time heart failure is the consequence of another disease, such as a heart attack, that has weakened the heart. Because heart failure is the end result of many different cardiovascular-related diseases, and because cardiovascular disease is so widespread, in the United States, 1 in 8 death certificates mention heart failure as a contributing cause.

Many genetic variants have been related to an increased risk of heart failure, often because the variants cause an abnormal amount of stress on the heart that can, over the years, cause the pump to wear out prematurely. In the United States, studies have indicated that the rate of heart failure is greater, and the age of onset younger among African Americans, although the cause of this is unknown. African Americans have between a 50 and 70 percent higher risk of heart failure between the ages of 45 and 64, and 1 in 100 will be afflicted by the disease by about the age of 40, which is 20 times the rate for the general population. Genetics holds the answer to this mystery.

A study published in 2002 in the *New England Journal of Medicine* investigated the association between variants in the *ADRA2C* and

ADRB1 genes and heart failure and found that individuals who had variants in both genes had a more than 500 percent increased risk of heart failure. The study also found that the frequency of these variants was much higher in people of African descent than in people of other ancestry; therefore, this may be one of the reasons heart failure occurs with so much more frequency in African Americans.

The *ADRA2C* and *ADRB1* genes are responsible for modulating the amounts of a hormone called norepinephrine, which is very similar to adrenaline. When a person has variants in both these genes, the heart is exposed to too much of this hormone and frequently beats too forcibly. Over many years, this causes the heart to be overstressed and overworked, which can eventually cause it to wear out and lose its ability to pump blood, thereby causing heart failure.

This discovery has direct implications for the identification and prevention of heart failure, because medications, such as beta-blockers, can undo the effects of these variants. If the medications are given before heart failure occurs, they could, presumably, prevent the progression of the disease and avoid heart failure altogether.

A number of genetic variants can cause the heart muscle to be overworked until it can no longer pump blood effectively.

In support of this theory, one follow-up study found that beta-blockers did provide significant protective benefit to the hearts of people who have variants in both of these genes. While additional studies are needed to confirm this result, it appears that this medication can be used to effectively outsmart the harmful variants and avert heart disease.

. . .

THE GENETIC BUTTERFLY EFFECT
(And the Heart Failure Genetic Variant Carried by More Than 65 Million People)

I'VE ALWAYS been amazed not only by how well organized our genetic code is but also by how even a little bit of chaos thrown into that code can affect a person's health. A change in just 1 out of the 6 billion letters in the entire genetic makeup of a person—that represents a change in 0.000000017 percent of the total—can lead to millions of people having an extremely harmful disease, such as heart failure. How can something so small cause something so significant in so many people? The answer is what I refer to as the genetic butterfly effect.

The term *butterfly effect* was coined by Edward Lorenz, an American mathematician, meteorologist, and MIT professor, who, in 1972, gave a talk titled "Predictability: Does the Flap of a Butterfly's Wings in Brazil Set Off a Tornado in Texas?" The premise, which exemplified an idea from chaos theory called sensitive dependence on initial conditions, was that very small variations in initial conditions, such as the flap of a butterfly's wings in Brazil, can eventually produce much larger effects, such as a tornado in Texas. Even though the first event seems inconsequential and the two events may be thought of as unrelated, they can be linked.

The genetic butterfly effect explains how a single, seemingly inconsequential change in the genetic makeup that we're born with can, 50 years later, lead to a life-threatening disease, such as heart failure, and how that single variant can also contribute to the leading cause of death for an entire continent.

While the incidence of heart failure is high in the United States, it is much higher in southern Asian countries where it is now the leading cause of death. In South Asia, heart failure most commonly affects people in the prime of life, at around the age of 45. It's been estimated that by 2030, heart failure will result in nearly 18 *million* years of lost productivity for South Asia and about 1.8 *million* years for the United States. If we are able to outsmart heart failure, people who would normally have been stricken with this disease in midlife will instead be able to live, work, and be productive for many more years.

In 2009, a genetic variant in the *MYBPC3* gene was confirmed to be associated with a substantially increased risk of heart disease and heart failure. This single variant is found in *1 in 100 people in the entire world*, and *1 in 25 people* in South Asia, which means it is in the genetic makeup of more than *65 million* people. The variant increases the risk of heart disease by approximately 500 percent if no instigating factors (such as high blood pressure or coronary artery disease) are present. If any of those factors do exist, the risk is increased even more. One study found that having one copy of the variant equates to almost a 90 percent chance that a person will experience heart disease within his or her lifetime. If instigating factors also exist or if the person's genetic makeup contains not one but two copies of the variant, the onset of heart disease can occur as early as 20 years old and the risk of sudden death is also substantially increased.

The *MYBPC3* gene is responsible for producing a protein that is essential for maintaining the structural integrity of the heart's muscle cells. The variant causes the gene to produce an abnormal, misshapen protein, which then disturbs the arrangement of other proteins within the heart muscle cell, leading to decreased structural integrity. The cell is unable to efficiently break down this abnormal protein so that, over many decades, the problem compounds because the disruptive protein accumulates throughout the heart, eventually leading to heart failure and death.

The first flap of the wings of this variant's genetic butterfly effect occurred about 30,000 years ago when it arose spontaneously in a distant ancestor in India. Back then, the average

continued

lifespan wasn't long enough for the variant to have time to create any significant health effects, so seemingly healthy parents passed it on to their children and then usually died of other diseases. Over thousands of years, the variant spread throughout all of South Asia. So, that single flap of the butterfly's wing has now been compounded to the point that its effect is felt by more than 65 million people.

Seeing the genetic butterfly effect in action provides a clear example of how predictive medicine can benefit not only individuals but entire populations. By working with the governments and charitable organizations of southern Asian countries, we can use the information we have about the *MYBPC3* variants to encourage the screening of entire populations and offer preventive services to people who are found to possess it. We can now disrupt the genetic butterfly effect so that entire populations are no longer destined for disease or death.

SOUNDING THE ALARM ON THE SILENT KILLER: HIGH BLOOD PRESSURE

High blood pressure, also referred to as hypertension, is, just as it sounds, a condition in which the pressure in your blood vessels is chronically elevated, causing damage to various organs. It is extremely prevalent, now affecting more than 1 *billion* people worldwide, with that figure projected to increase by 60 percent within the next 15 years. In the United States, this equates to 1 in 3 adults and it's even higher in Europe, Asia, and Africa. In more than 65 percent of those who have it, hypertension is uncontrolled, and their blood pressure remains high on a regular basis.

High blood pressure is often referred to as the silent killer because it has very few symptoms, and most people don't even know they have it until, over the course of many years, it causes widespread damage to many different organs, most notably the heart and the brain. Because of this, hypertension contributes significantly to the risk of heart attack, heart failure, stroke, and brain aneurysm.

Although some rare forms of hypertension—generally disorders that prevent the kidneys from properly eliminating salt from the body—are predominantly due to just a single genetic variant, the majority of people have hypertension as the result of many different genetic and nongenetic factors. Nongenetic factors that increase the risk of hypertension include obesity, a sedentary lifestyle, alcohol intake, sodium consumption, and even vitamin D deficiency. Therefore, blood pressure can be significantly altered with lifestyle modifications, and this means it is an excellent disease to predict on a genetic level and prevent through lifestyle choices.

Genetic screening is also a way to sound the alarm on this silent killer by alerting people who are genetically predisposed to the disease to have their blood pressure checked at least once a year by a healthcare provider and even to purchase a home blood pressure monitor. This increased surveillance will help detect previously silent hypertension, allowing for immediate intervention and treatment.

Among the variants found to be associated with high blood pressure are those in the *AGT* gene, which plays a role both in dictating the constriction of blood vessels and in modulating the amount of sodium in the blood. When variants cause the gene to malfunction, it becomes overactive, leading to increased blood vessel constriction and higher levels of sodium in the blood, both of which lead to higher blood pressure. Women who have these variants are also at increased risk for developing hypertension during pregnancy.

Genetic screening may identify the cause of high blood pressure so that treatment can be specifically targeted to the cause.

Tens of thousands of years ago, salt, which is now all too prevalent in our diet, was a rare and cherished resource. When salt was scarce, people were at risk for electrolyte imbalances, which contributed significantly to illness and even death. Researchers have proposed that during human evolution variants within genes involved in sodium retention, such as the *AGT* gene, may have provided an evolutionary advantage because they enabled people to retain more salt. This hypothesis is supported by the fact that many of these variants are prevalent throughout the world, indicating that they would, at some point, have conferred a survival advantage. Now, however, with most people consuming too much salt, these same sodium retention variants are contributing to hypertension.

As a frame of reference, the average American adult consumes about 3.5 grams of sodium per day, while most health organizations recommend no more than 2.3 grams per day and no more than 1.5 grams per day for people who have hypertension. Studies focusing on the *AGT* genetic variants found that for people with these variants, reducing sodium intake lowered the risk of hypertension enough to *also* lower their risk of coronary artery disease and stroke. For people who did *not* have those variants, however, reducing sodium intake did *not* significantly lower the risk of hypertension. This is consistent with medical dogma stating that sometimes hypertension is salt sensitive and sometimes it's not.

Dietary Approaches to Stop Hypertension (known as the DASH diet) is promoted by the National Institutes of Health to decrease high blood pressure. The primary components of the diet are a reduction in both sodium and fat intake and an increased consumption of fruits and vegetables. One study that looked at this diet found that it was effective in reducing hypertension in people who had *AGT* variants. It's been known for a while that African Americans respond better than most to this diet, and that may be because some AGT variants are found much more frequently in African Americans than in other ethnicities.

The DASH diet has been successful in treating high blood pressure caused by specific genetic variants.

GENETICALLY TAILORED PREVENTION

As we've just seen with the *AGT* variants and high blood pressure, not only diseases and conditions but also the preventive interventions that are most likely to work for a particular individual are, at least in part, dictated by genetics. Predictive medicine takes this into account so that many of the recommended preventive measures can be personalized to each person's genetic makeup.

GENETICALLY TAILORED LIFESTYLES FOR CARDIOVASCULAR HEALTH

Many physicians have become disillusioned with the efficacy of providing lifestyle modification advice to their patients because they have come to believe that these recommendations have little impact on disease prevention and treatment. Therefore, many doctors choose to forgo lifestyle recommendations and start medication treatment right away. Research has shown, however, that recommendations about changes in lifestyle patterns, such as changing one's diet, actually do have some beneficial effect. I believe that the problem primarily results from the fact that most lifestyle recommendations don't have personal meaning to most patients and therefore lack impact. One way this can be overcome is if the recommendations are tailored to a person's genetic code.

In addition to a link between low-sodium diets and *AGT* variants for high blood pressure, a growing number of other interactions have been found. For example, preliminary studies have identified variants in several genes that increase a person's risk of stroke, but *only* if the person smokes cigarettes.

Even the degree to which stress has a detrimental effect on our cardiovascular system appears to be determined by our genes. As Jerry Seinfeld once said, highlighting how stressful public speaking is for many people, "The average person at a funeral would rather be in the casket than doing the eulogy." One preliminary study looking at variants in the *ADRB1* gene—the same gene we discussed in relation to heart failure—asked people with coronary artery disease to speak publicly and then measured the amount of oxygen going to the speaker's heart. The study found that when people with variants in the *ADRB1* gene were stressed, such as by having to speak publicly, the blood flow to the heart was severely diminished, meaning that people with these variants may be at much higher risk of heart attacks when exposed to stressful situations. So even a preventive recommendation to decrease stress levels, such as by avoiding stressful situations and practicing yoga, can be tailored to an individual's genetic makeup.

GENETICALLY TAILORED MEDICATIONS

Oliver Wendell Holmes Sr., one of the most astute physicians of the 19th century, was extremely discouraged by the ineffectiveness of most medications. While discussing the subject, he famously commented that if all medications "could be sunk to the bottom of the sea, it would be so much the better for mankind—and all the worse for the fishes!"

The pharmaceutical industry has come a long way since the 1800s, having produced life-saving vaccines and treatments for a wide variety of diseases that were previously untreatable. However, many medications are effective for only about 60 percent of people. And worse, adverse reactions to medication among people in the United States are responsible for more than 100,000 deaths each year. Both of these disturbing realities—poor effectiveness rates and adverse reactions—are primarily due to variations in our individual genetic makeup. In terms of heart disease, no medication better exemplifies this than one of the most prescribed medications in the world: statins.

Statins are widely prescribed to lower cholesterol levels, but for some people—particularly those with variants in the *SLCO1B1* gene—they not only are ineffective but also are associated with adverse reactions such as muscle pain. The *SLCO1B1* gene is important for regulating how much of a drug enters the bloodstream. Variants in this gene appear to alter its function, causing more of the drug to circulate in the blood, which exposes the body to higher doses and increases the risk of adverse reactions. A recent study found that variants in this gene occur in about 15 percent of people and are responsible for 60 percent of the most common adverse reactions to statins. Knowing this, we can now anticipate the likelihood of a patient's having an adverse reaction to statins and either prescribe a different class of medication or start at a lower dose.

In addition to statins, the effectiveness of many other medications used to treat cardiovascular disease, including Plavix and beta-blockers, have been associated with particular genetic variants. Even aspirin, which is used to thin the blood and decrease the risk of blood clots, heart attacks, and strokes, isn't effective for people who have specific variants.

Predictive medicine can now provide very useful information about the effectiveness and risk of side effects of medications used to prevent and treat cardiovascular disease.

The final example of how predictive medicine can improve the use of medications has to do with the blood-thinning drug warfarin (Coumadin), which is one of the drugs most commonly prescribed throughout the world to prevent heart attacks, strokes, blood clots, and other life-threatening illnesses. Determining the initial dose can be difficult because a number of variants influence how people respond to the drug. Some respond normally, meaning that their blood thins as expected when they are given the starting dose. Others are very sensitive to the medication, and this causes their blood to become too thin, which puts them at high risk of life-threatening bleeding, particularly into the brain. And still other people are resistant to warfarin, so the starting dose doesn't thin their blood enough, which means they are not protected against heart attacks and strokes even while taking the medication. All of these possibilities can be avoided, however, because genetic screening is able to determine not only who will be sensitive and who will be resistant but also the precise starting dose that will be most effective for the person. As discussed in Part I, a senior member of the FDA's economic staff concluded that if genetic screening were used prior to prescribing this medication, 85,000 serious bleeds, 17,000 strokes, and other warfarin-associated adverse reactions could be avoided each year, resulting in a net healthcare savings of as much as $1 *billion* per year.

Examples of Cardiovascular-Related Panels

Cardiovascular Panel

- High cholesterol and triglyceride levels
- High blood pressure
- Heart attacks
- Strokes
- Coronary artery disease
- Heart arrhythmias and causes of sudden death (including medication-induced arrhythmias)
- Cardiomyopathies
- Heart failure
- Blood clots
- Brain aneurysms
- Aortic aneurysms
- Effectiveness, dosing, and adverse reactions to cardiovascular-related medications
- Effects of specific diets and exercise on cardiovascular disease risk
- Effects of smoking on cardiovascular disease risk
- Effects of stress on risk of worsening cardiovascular disease
- Nicotine addiction risk
- Alcohol abuse risk
- Biomarker levels (including CRP, homocysteine, adiponectin, *p*-selectin, and IL18)
- Anemia and other abnormalities of the blood
- Peripheral arterial disease
- Structural heart defects
- Marfan's syndrome
- Bleeding disorders (such as hemophilia)
- Kawasaki's disease

- Periodontitis
- Obesity
- Type 2 diabetes

For additional panels related to cardiovascular health, please visit www.OutsmartYourGenes.com/Panels.

9

A New Strategy in Our War Against Alzheimer's Disease

MISCONCEPTION: If Alzheimer's disease is my fate, I'd rather not know because there's nothing I can do about it anyway.

FACT: The most powerful weapon you have against Alzheimer's disease, even decades before it strikes, is knowing that you're at risk, because you *can* substantially lower your risk of this disease.

Everyone has so-called senior moments. As we age, it's normal for us to forget a phrase or the names of people we seldom see; however, problems recalling the names of good friends and family members, or difficulty performing a normal activity of daily living, such as balancing your checkbook (assuming you were able to balance it before), may signal a serious problem. While living to a ripe old age is a dream many of us share, we also share the dread that old age might be accompanied by dementia.

Dementia is defined as a decline in memory and cognition, such as the ability to process and learn information, recognize or identify objects, communicate, and make sound judgments. Dementia is what people used to call senility—when grandma became feebleminded—

and for years it was assumed that senility was a normal part of aging. Today, we consider dementia a disease requiring treatment.

Alzheimer's disease is responsible for approximately 70 percent of cases of dementia. Because it is so prevalent throughout the world, almost everyone is familiar with the symptoms of Alzheimer's disease, which usually start with memory loss that deteriorates significantly over time and is then accompanied by confusion, impaired judgment, and disorientation. Eventually the Alzheimer's patient cannot recognize loved ones and has trouble speaking, walking, and eating, all of which result in the person's being bedridden and ultimately leads to death.

There is currently no cure for Alzheimer's disease. Once the disease has taken hold, the only thing a physician can do is prescribe medications that slow down the inevitable. Eventually, everyone with Alzheimer's deteriorates and dies. This is why enacting preventive measures that lower a person's risk of getting Alzheimer's is the only viable option we currently have to fight back against the disease.

PROTECTING THE SPARK OF LIFE

When it comes to genetic screening, Alzheimer's disease has received a bad rap. All too often I've heard people say, "I want to know my risk for everything . . . except Alzheimer's disease." The primary reason for this exclusion is that many people believe we aren't able to modify our risk for Alzheimer's, but that is not true. By instituting preventive measures as early as possible, we can significantly lower our risk of contracting the disease so that it either never occurs or starts much later in life and progresses more slowly.

If you know you're genetically predisposed to Alzheimer's disease, you can enact preventive measures that will decrease your risk.

Many people say that they'd rather go quickly with a heart attack than slowly deteriorate with Alzheimer's disease. But just because we fear something doesn't mean that the best way to cope is by putting our heads in the sand. On the contrary, the way we are going to defeat this devastating disease is by taking lifelong action against it.

UNDERSTANDING ALZHEIMER'S

Alzheimer's disease is a debilitating neurological disorder that usually afflicts people late in life—that is, after age 65. It is characterized by the formation of abnormal deposits in the brain called plaques and tangles, which affect the neurons and diminish the effectiveness of the brain's electrical activity. As the deposits accumulate, nerve cell connections (called synapses) are reduced, the cells eventually die, and the brain is unable to function properly. Areas of the brain that influence short-term memory tend to be affected first. Later, the disease works its way into parts of the brain that control other intellectual and physical functions. Death usually occurs within 8 years of diagnosis.

THE ALZHEIMER'S WAR

In 1907, when the disease was first described by Dr. Alois Alzheimer, the average life expectancy even in developed nations was low enough so that most people died from another cause before Alzheimer's could manifest. Now, however, with the significant increase in our life expectancy over the last century, the number of Alzheimer's cases is increasing rapidly. In 2010, the first baby boomers reached their 65th birthday and by 2029 all baby boomers will be 65 years old or older, with 70 million people over the age of 65 in the United States alone. The disease currently affects *1 out of every 8* people over the age of 65 and nearly *1 out of 2* over the age of 85 in the United States. Worldwide, more than 26 million people have Alzheimer's disease, and unless science finds a way to intervene, this number is expected to quadruple within the next 40 years to the point at which *1 in every 85* people throughout the world will be affected.

The victims of Alzheimer's disease are not just those who are afflicted but also their families, loved ones, and community. Nearly 10 million friends and family members in the United States, including 250,000 children between the ages of 8 and 18, currently act as caregivers to someone with Alzheimer's disease, providing more than 8.5 *billion* hours of unpaid care each year. More than half of these caregivers report feeling that they are on duty 24 hours a day. And due to the long-term, progressively declining nature of the disease, which creates a tremendous amount of psychological distress for the caregiver, *2 out of 3* caregivers experience some degree of depression.

The bereavement process with Alzheimer's is unique because loved ones often begin the mourning process while the patient is still physically alive but otherwise already lost to them. And along with all the other stresses, many family members also fear that they might inherit the disease.

As you might imagine, the costs of Alzheimer's disease are staggering: about $150 billion a year in the United States. Medicare spends three times as much on Alzheimer's as any other disease and 50 percent of all nursing home costs are related to dementia and Alzheimer's.

Our entire civilization is fighting a war against Alzheimer's, and right now the disease is winning. The statistics are certainly alarming, especially because the numbers are getting worse. The word *alarming* is appropriate because these statistics should alert us all to the true gravity of the situation we now face and the need to launch a more aggressive counteroffensive.

KNOW THINE ENEMY

The more common, sporadic form of Alzheimer's disease, which is responsible for 95 percent of all cases, is about 70 percent genetic and 30 percent nongenetic in origin. The fact that this form of the disease isn't fully dependent on our genes means that we *can* manipulate the nongenetic factors to decrease our risk. In the future we may be able to use gene therapy or genetic engineering to change our genetic makeup, but for now we must use the tools we have, and those center on decreasing risk by modifying those nongenetic factors.

The rare, familial form of Alzheimer's disease is almost 100 percent caused by genetic factors, and there is no way to lower the risk for this form of the disease. However, detecting the variant that causes it allows you to use family planning options to stop it from affecting future generations. For the remainder of this chapter, however, we'll be focusing solely on the more common form of Alzheimer's.

THE *APOE* GENETIC VARIANTS

The *APOE* gene produces a protein that has a number of functions in the brain, but three of its most important ones are helping break down harmful deposits, assisting in the maintenance and repair of the structural integrity of brain cells, and maintaining something called synaptic plasticity. Synaptic plasticity enables neurons to communicate with one another, and maintaining synaptic plasticity is vital for learning and memory. If the *APOE* gene doesn't function properly, harmful deposits can build up, the structural integrity of brain cells is compromised, and synaptic plasticity may deteriorate. Therefore, the *APOE* gene is essential for maintaining brain health and is the most important gene involved in Alzheimer's disease.

The *APOE* gene has three primary variants: *E2*, *E3*, and *E4*. When it contains the *E3* variant it is able to produce a normal amount of its protein, when it contains *E2* it is able to produce a higher amount of its protein, but when it contains the *E4* variant the gene doesn't function as well and produces low levels of its protein. And what science has figured out concerning the relationship between Alzheimer's and these three *APOE* versions is basically this: *E2* is protective, *E3* is neutral, and *E4* is harmful. Because we all contain two copies of the *APOE* gene, we can have two copies of *E2* (*E2/E2*), one *E2* and one *E3* (*E2/E3*), one *E2* and one *E4* (*E2/E4*), and so forth.

In terms of its effects, *E4* always outweighs all other variants. No matter what the other copy of your *APOE* variant is, you're still at increased risk of Alzheimer's. But if you have *E2/E2* or *E2/E3*, you may actually have a much *lower* risk of Alzheimer's disease. So determining the particular variants in your *APOE* gene may actually let you know that you are at *lower risk* for Alzheimer's than the general population. The following chart indicates the risk statistics associated with each of these variants as they apply to Caucasians.

ALZHEIMER'S DISEASE AND *APOE*	
APOE VARIANTS	RISK
E2/E2 and *E2/E3*	40 percent lower
E3/E3	No increased risk on its own, but risk may be increased if variants also exist in the *TOMM40* gene
E2/E4	160 percent increased
E3/E4	220 percent increased
E4/E4	> 1,000 percent increased

Asians (Japanese, Chinese, and Koreans) have an even greater increase in risk of Alzheimer's disease if their genetic makeup contains one or two copies of *E4*, while blacks and Hispanics have far less of an increase.

The *E4* version of *APOE* may account for up to 50 percent of all Alzheimer's disease cases. A study conducted in 2009 by a team of doctors at the Mayo Clinic found that memory decline in people with *E4* actually started *before* the age of 60 and accelerated at a faster pace compared to people without *E4*. This supports other studies that found an earlier and faster decline in memory and learning ability for *E4* carriers. Therefore, *E4* not only greatly increases a person's risk of Alzheimer's but also causes the disease to manifest earlier in life.

In the near future, we might be able to provide gene therapy that represses the harmful effects of *E4*, or we may even be able to provide genetic engineering that changes a person's *E4* to *E2*. Until that time, however, we can use its presence to predict our risk of future disease and, with that foreknowledge, start preventive strategies as early as possible.

ALZHEIMER'S RISK: OTHER GENETIC VARIANTS

The *APOE* gene is not the only one associated with Alzheimer's risk. Another gene that affects risk is *SORL1*, which is involved in protecting the brain against the formation of the senile plaques that, along with tangles, are the primary changes observed in the brain. Variants within *SORL1* diminish its protective abilities and increase a person's risk of Alzheimer's. The gene, however, also contains other variants that appear to boost its function so that it works better, and these variants are associated with a *decreased* risk of the disease.

While the *SORL1* gene modifies a person's risk of Alzheimer's regardless of whether he or she has the *E4* variant of *APOE*, variants within *GAB2* appear to increase the rate at which harmful deposits form *only* in the brains of people with *E4*. When a person's genetic code contains both *E4* and specific *GAB2* variants, his or her risk of Alzheimer's disease increases substantially more than it would if the person had only the *E4* variant.

In addition, it now turns out that the seemingly neutral *E3* variant of *APOE* is not so benign after all because it is associated with an increased risk of Alzheimer's disease in a certain subset of people. In 2009, the director of Duke University's Drug Discovery Institute found that variants within the *TOMM40* gene could be used to predict not only increased risk of Alzheimer's disease but also the age of onset in people who also possess the *E3* variant of *APOE*. The *TOMM40* gene is important for the normal functioning of the mitochondria, the powerhouses of cells. Variants within this gene impair its ability to function normally, leading to mitochondrial dysfunction and the accumulation of damage within brain cells. This, in turn, leads to the premature brain cell death that occurs in Alzheimer's disease.

Preliminary research has found that variants in the *TOMM40* gene may be very useful in predicting an additional 35 percent of a person's genetic risk of Alzheimer's. This means that by combining information on variants within the *APOE*, *SORL1*, *GAB2*, *TOMM40*, and other genes, doctors may be able to predict up to 90 percent of a person's genetic risk of Alzheimer's. Although the *TOMM40* data are currently being confirmed, they may signify a potential leap forward in our ability to predict Alzheimer's disease.

Clearly, a comprehensive analysis of the genome is required if we are to take into account all the variants that may be associated with Alzheimer's. The goal of predictive medicine, however, is not just to identify increased risk but also to institute personalized prevention for those who need it most.

Predictive medicine can provide personalized prevention of Alzheimer's disease.

GENETICALLY TAILORED PREVENTION

By instituting a number of preventive measures, you can decrease your risk of Alzheimer's disease. And as with other diseases, this prevention can be genetically tailored so that it is most effective for you.

One of the primary reasons some people opt out of genetic screening for Alzheimer's disease is that there is no cure, and treatment options are both limited and minimally effective. Delaying or preventing Alzheimer's *before it manifests* is the best strategy we currently have for combating the disease.

Even though Alzheimer's strikes the majority of people after age 65, prevention, in my opinion, can begin at birth and continue throughout one's life. Because it is an insidious disease that causes significant, irreversible damage long before the first symptoms even appear, the diagnosis may not occur until age 65 or later, but harmful changes to the brain almost always start to occur up to 20 to 30 years before that. And nongenetic instigating factors encountered as early as childhood also appear to influence the risk of Alzheimer's.

Delaying or preventing Alzheimer's before it manifests is the best strategy for combating the disease.

Therefore, the battle for the long-term health of our brain starts not at midlife but when we are children.

BUILD AND MAINTAIN A COGNITIVE RESERVE

Building a large cognitive reserve throughout life is a very proactive way of decreasing your risk of Alzheimer's disease. Think of your brain as a reservoir filled not with water but with neurons. When your brain's neuron reservoir dips below a certain level because neurons either die or become incapacitated, Alzheimer's occurs. Therefore, Alzheimer's can be thought of as a drought in the brain. Even though the reservoir level may have been slowly decreasing for decades, it isn't until the level dips below a certain threshold that the drought becomes noticeable and symptoms of Alzheimer's appear. And the lower the reservoir gets, the worse the symptoms become.

Using this analogy, the *E4* variant of *APOE* can be thought of as a large crack in the foundation of the reservoir that allows the brain to lose neurons more quickly than it would otherwise, which means that people with this variant reach the Alzheimer's threshold at a younger age and, therefore, start to experience symptoms of memory loss and learning difficulties earlier than those who don't have the variant. This process is similar to what occurs in people with variants in their *TOMM40* gene, except that the crack in their foundation isn't as large and, therefore, their loss of neurons is not as rapid. Luckily, however, the reservoir of neurons in your brain can be increased substantially throughout life by creating a reserve. The term *cognitive reserve* refers to this concept—a *reserve* of neurons that protects against Alzheimer's disease. This reserve effectively increases the neurons in your reservoir so that even if you do have a leak, such as occurs with *E4*, the additional

Protect yourself against Alzheimer's disease by building and maintaining a cognitive reserve throughout your life.

neurons in your reservoir can ensure that the level remains above the disease threshold for a much longer period of time.

There are many ways to stimulate your brain to create a cognitive reserve. The first and foremost of these is education. Numerous studies have found that there's an inverse correlation between the number of years of education and a person's risk of Alzheimer's disease. The risk of dementia and Alzheimer's disease is about 15 percent lower for people with 12 to 15 years of education, and 35 percent lower for people with more than 15 years, as compared with those who have fewer than 12 years of schooling.

But even if you are past school age, you needn't fret, because you can still increase your cognitive reserve by exercising your mind on a regular basis (called cognitive training) in much the same way you'd exercise your body to maintain physical fitness. Cognitive training has been shown to stimulate the growth of neurons and increase synaptic plasticity. Mental exercises, which can be performed three or more times a week for about an hour each time, include playing bridge and other card games, playing chess, solving puzzles (such as crossword puzzles or sudoku), attending lectures, learning a new language or skill, and starting a new hobby. And the best types of mental exercises appear to be those that involve social interaction, because building a strong social network of friends has also been associated with a protective effect against Alzheimer's, possibly because it helps alleviate stress.

Currently, studies are under way to determine whether using modern technologies, such as playing video games, can also help build cognitive reserves and protect against Alzheimer's disease. Preliminary research is promising, showing that puzzle-based video games with increasing levels of difficulty can significantly improve memory and synaptic plasticity. If further research indicates that it is indeed beneficial, perhaps parents will start *encouraging* their children to play even more video games (and parents and grandparents might start playing them too).

Although building a cognitive reserve hasn't yet been shown to prevent the plaques and tangles of Alzheimer's disease from building up in the brain, it does significantly *delay* the onset of symptoms. And since Alzheimer's usually afflicts people when they are 65 or older, delaying the onset of symptoms by one or two decades will allow for considerably more of the golden years to be spent enjoying life to the fullest. Furthermore, because we may very well succumb to some other illness by the time we reach our 80s or 90s, it would seem that the longer we are able to delay the onset of Alzheimer's, the less likely it is to have a significant effect on our lives.

In Chapter 7 we discussed the association between head trauma and a substantially increased risk of developing Alzheimer's for those with the *E4 APOE* variant. For people with this variant, a head injury is comparable to smashing at the wall of the reservoir with a sledgehammer. Normally the wall is strong enough to withstand such a barrage, but in people with the *E4* variant, the reservoir already contains a large crack, which head trauma significantly increases; this, in turn, increases the rate at which neurons are lost from the reservoir. Researchers are currently working on a medication that will help protect against cognitive decline after head injury in people with *E4*, but while there's been success in studies on mice, the drug has not yet been tested on humans. Therefore, at this time, the best line of defense for people with this variant remains the avoidance of head trauma.

WHAT'S GOOD FOR THE HEART IS GOOD FOR THE HEAD

As it turns out, almost all nongenetic factors and even some genetic variants that increase your risk of heart disease also increase your risk of Alzheimer's. Although that doesn't sound good, it can actually be beneficial because it means you'll receive double the benefit from the same preventive strategies. For example, heart-unhealthy factors, such as smoking, being overweight or obese, or having high cholesterol levels, are also associated with an increased risk of Alzheimer's disease, so overcoming any of these will substantially reduce your risk of both conditions.

Physical activity appears to have multiple effects on your brain that go far beyond cardiovascular health. It not only increases synaptic plasticity but also stimulates the growth of neurons, thereby increasing your cognitive reserve. In addition, in studies of laboratory animals, physical exercise appears to stop the formation of Alzheimer's-related plaques in the brain.

For those who lead a sedentary life, the risk of Alzheimer's disease is more than 200 percent greater than it is for those who are physically active. For people with *E4*, being sedentary increases the risk of Alzheimer's even more, but regular exercise has been shown to reduce that risk significantly.

The greatest benefit from physical activity most likely comes from either 20 minutes of vigorous exercise three times a week or moderate-intensity exercise for 30 minutes five times a week. And the best forms of physical exercise appear to be those that get the heart pumping, such as swimming, jogging, biking, or playing tennis. In Chapter 5, we talked about the fact that the physical activity you do at work or when commuting to work doesn't appear to be anywhere near as effective as leisure-time physical activity for slowing down your genetic clock. And the same appears to be true for helping to lower your risk for Alzheimer's disease—to be beneficial the physical activity has to be completed during leisure time.

Heart-healthy diets seem to be brain healthy as well. There is already a substantial body of research supporting the ability of the Mediterranean diet to protect not only against cardiovascular disease and cancer but also against diseases that affect the brain, including Alzheimer's and Parkinson's diseases. A Mediterranean diet includes a lot of extra-virgin olive oil, vegetables, nuts, unrefined grains, fish, and fresh fruits along with small quantities of meat, dairy, and refined sugars. The most likely

Leisure-time physical activity is not only good for your heart but also protects you against Alzheimer's disease.

reason this diet decreases the risk of Alzheimer's is that it is high in natural antioxidants and quite low in calories and saturated fats.

SEVERAL CUPS OF COFFEE A DAY
MAY KEEP ALZHEIMER'S AWAY

AS WE'VE discussed, caffeine is one of the most widely consumed substances in the world. However, it may have more beneficial effects than just allowing you to get through the morning. Research now indicates that caffeine consumption actually *lowers* a person's risk of Alzheimer's. And not just any caffeine consumption but a considerable amount every day. In fact, people who drink tea and cola see little benefit due to the lower amounts of caffeine in those beverages. The greatest protection goes to avid coffee and espresso drinkers.

When you drink something with caffeine in it, that caffeine goes straight to your brain and interacts with brains cells in the same way as a medication. Specifically, caffeine temporarily *turns off* parts of the brain that *slow down* the speed at which the brain works. If you turn off something that's slowing you down, you speed up, and that's pretty much how caffeine works.

In addition to stimulating the brain and increasing cognition and memory, caffeine may also help reduce inflammation throughout the brain. Chronic inflammation causes the body a tremendous amount of harm, and substances that reduce inflammation, such as aspirin, red wine, and caffeine, can be helpful in protecting the body against disease. Because caffeine appears to help decrease inflammation in the brain, it's thought of as being neuroprotective.

In addition to caffeine, coffee also contains other beneficial substances such as polyphenols. We discussed these in the last chapter when we talked about the cardiovascular benefits of red wine. The polyphenols in coffee, just like those in red wine, act as antioxidants and protect the brain against damage. In fact, studies indicate that polyphenols lower a person's risk of Alzheimer's disease.

The length of time you've been drinking your coffee is also significant, because its health benefits aren't realized until it's consumed on a regular basis for two months or longer. Although studies indicate that short-term use of coffee augments attention and alertness, its positive effects on memory and cognition appear to occur only with long-term use.

While confirmatory research is necessary, studies have found that drinking three to four cups of coffee or two to three espressos a day may lower Alzheimer's disease risk by about 50 percent. And for those individuals with the high-risk *E4* version of *APOE*, coffee or espresso appears to have an even greater protective effect, reducing the risk by around 70 percent.

Before you start brewing a pot of coffee, however, keep in mind that copious amounts of coffee are not always beneficial to health. Too much can cause restlessness, anxiety, and irritability. And, as with any medication, whether you're likely to experience these adverse reactions is based on your specific genetic makeup.

The daily consumption of beverages other than coffee and espresso also has an effect on Alzheimer's risk. Drinking one to two glasses of red wine not only benefits cardiovascular health but also appears to decrease the risk of Alzheimer's disease. And studies suggest that red wine may be the *only* alcoholic beverage that decreases the risk of Alzheimer's. However, unlike coffee and espresso, wine consumption appears to decrease a person's risk of Alzheimer's only if he or she *does not* have the E4 variant. The reason for this is most likely that the negative consequences of E4 overwhelm any benefit obtainable from wine.

PREVENTION IN A PILL

The scientific literature consistently identifies two classes of medication—statins and nonsteroidal anti-inflammatory drugs (NSAIDs)—as lowering the risk of Alzheimer's. The most common NSAIDs are aspirin, ibuprofen (such as Advil and Motrin), naproxen (Naprosyn), and diclofenac (Voltaren). Aspirin is a unique NSAID because, unlike ibuprofen or naproxen, it thins the blood (makes platelets less sticky), which is one of the reasons it is used to prevent blood clots, heart attacks, and strokes. Tylenol (acetaminophen or paracetamol) is sometimes thought of as being part of this group, but it is technically *not* an NSAID, and studies have shown that it does not appear to be effective in reducing the risk of Alzheimer's.

The benefit of NSAIDs in relation to Alzheimer's prevention appears to occur because they not only reduce inflammation throughout the body but also may directly prevent the formation of Alzheimer's-related plaques in the brain. Some of the first evidence that NSAIDs may lower a person's risk of Alzheimer's came from studies in the early 1990s that showed an inverse relationship between arthritis and dementia. People with arthritis had an almost 50 percent lower risk of Alzheimer's disease than those who did not have arthritis, and this left researchers wondering, "What is it about people with arthritis that could be causing them to have a lower risk of Alzheimer's disease?" It quickly became apparent that the one type of medication taken by almost all people with arthritis on a daily, long-term basis is an NSAID, because NSAIDs help reduce both the pain and the inflammation in their arthritis-ridden joints. Because of this, researchers set up studies to determine whether NSAIDs were actually responsible for the observed lower risk of Alzheimer's and, if so, whether they could be used for prevention.

A large number of studies have investigated the association between NSAIDs and Alzheimer's risk, with the majority indicating that daily use of NSAIDs for longer than 2 years is beneficial. And research conducted within the past few years is now providing the first evidence that NSAIDs are primarily effective for people who have one or two copies of the *E4* version of *APOE*, reducing their risk by about 60 percent. Studies of laboratory animals have shown *E4* to be associated with increased inflammation in the brain, and NSAIDs directly counteract this. Because of these findings, I recommend daily NSAIDs to many of my patients who have one or two copies of *E4*. However, since aspirin increases the risk of stomach ulcers and bleeding, it is important to always consult with your personal physician before using it for the prevention of any disease.

We've already discussed statins in relation to cardiovascular disease. Because Alzheimer's-related plaque accumulation in the brain is increased when a person has high cholesterol levels, statins, which normalize cholesterol levels, also help slow down plaque formation. And statins don't just reduce the risk of Alzheimer's by lowering cholesterol; they also appear both to inhibit the formation of Alzheimer's-related tangles and to decrease inflammation.

Medications such as NSAIDs and statins appear to be very effective in lowering the risk of Alzheimer's.

When it comes to the brain, however, not all statins are equal: Some, like pravastatin (Pravachol), have difficulty entering the brain from the blood, but others, like simvastatin (Zocor) and lovastatin (Mevacor) can. Research also suggests that statins must be started when the person is still young if they are to be effective for prevention.

Although more studies are needed, the data concerning the benefits of statins in reducing Alzheimer's risk are convincing enough for me to recommend prophylactic brain-penetrating statins to patients starting in their late 30s to early 40s if they are genetically predisposed to the disease. Because the adverse reactions associated with statin therapy can be mitigated by tailoring the starting dose of the medication to a person's genetic makeup and because the potential benefits are so high, I believe this is a good preventive strategy to use until additional studies are conducted. In our war against Alzheimer's we do not have the luxury of taking a wait-and-see approach because preventive strategies appear to be most beneficial when started at a younger age.

VITAMIN E-RRONEOUS

A KEY component of fighting against a disease such as Alzheimer's is making sure you don't waste your time or money on prevention that doesn't work. Fortunately, researchers around the world are diligently studying the effectiveness of preventive strategies. When one particular prevention starts to get a lot of public attention, this frequently attracts the attention of researchers, who then conduct studies to test whether all the hoopla is valid. One of the more recent therapies to receive this kind of attention was taking supplemental vitamin E for Alzheimer's prevention.

Because of its purported health benefits, people were popping vitamin E pills like, well, vitamins for the prevention of everything from Alzheimer's and cardiovascular disease to cancer. However, many of the initial studies have found no benefit and sometimes even harm from taking vitamin E supplements.

In 2005, researchers at the Johns Hopkins School of Medicine, along with a multinational research team, published a massive study that analyzed the results of 19 clinical trials of vitamin E involving a total of more than 135,000 people. The researchers found that taking 150 IU or more of vitamin E per day actually *increased* the risk of death in general. Compare this to the 14 IU that the average person consumes from food each day and it's evident that 150 IU may be way more than evolution designed our bodies to process. The study also found that the higher the dose, the greater the risk, with the risk of death increasing even more for people who take 400 IU or more per day. Further studies also showed that vitamin E supplementation did not prevent Alzheimer's, cardiovascular disease, or cancer and instead only appeared to increase risk of death. This means that the megadoses of vitamin E people commonly take for prevention may actually be mega-bad.

It appears that taking high doses of vitamin E may throw off the body's natural antioxidant balance, thereby increasing rather than reducing oxidative damage. Another study found that vitamin E may disrupt the body's ability to neutralize toxins, leading to a potential buildup of harmful substances.

Just because a substance has the seemingly innocuous word *vitamin* in its name doesn't necessarily mean it's good for you. If you have proper nourishment from food, you're probably getting all the vitamins you really need unless blood tests or your genetic makeup indicates otherwise. To outsmart something as dastardly as Alzheimer's, we cannot afford to waste our time on prevention that doesn't work. Although it was once thought to be helpful, vitamin E is now thought to offer nothing but false hope and, according to a number of studies, an even greater risk of death.

COMPLEMENTARY PREVENTION

In North America and Europe, the term *alternative medicine* is used to refer to nontraditional healthcare options, including treatments and interventions such as acupuncture, yoga, meditation, supplements, and herbology. When these techniques are used in conjunction with traditional Western medicine, the alternative methods are generally referred to as *complementary medicine*. Complementary preventive strategies have been shown to be effective in decreasing the risk of Alzheimer's and should be considered another powerful weapon in our arsenal.

Among the most popular supplements used for the prevention of Alzheimer's are omega-3 fatty acids. These are essential fatty acids, which means that they cannot be manufactured by the body and must, therefore, be obtained from food or in supplements. The most beneficial forms of omega-3s (known as DHA and EPA) come primarily from fish. They appear to have significant effects throughout the body and have been shown to lower cholesterol and triglyceride levels, reduce inflammation, protect against heart attacks, and have sizable benefits for overall health. Large studies have repeatedly shown that omega-3s can reduce the risk of cardiovascular disease by about 40 percent. In fact, the benefits of omega-3s are so apparent that in 2000 the FDA officially gave "qualified health claim" status to their ability to reduce the risk of coronary artery disease.

Omega-3s are found in high quantities in the membranes of brain cells and are required for the cells' normal functioning. Because of this, as well as their ability to modulate cholesterol levels and reduce inflammation, omega-3s have been widely investigated in relation to the prevention of Alzheimer's. Studies have found that low levels of omega-3s, such as if a person doesn't eat fatty fish more than three times per week, are associated with an increased risk of Alzheimer's and that this risk can be reversed if consumption is increased. The studies found that, in addition to their other benefits, omega-3s also appear to have a direct impact both on inhibiting the formation of plaques and tangles and on increasing the ability of brain cells to survive, thereby slowing down the rate at which neurons are lost from the reservoir.

Omega-3s from fish or fish oil pills appear to decrease the risk of Alzheimer's by about 40 percent. However, a growing number of studies that take into account the *APOE* gene have found that people with *E4* actually have far less, if any, benefit from omega-3s. *E4* and fish oil may not go well together because *E4* is associated with producing an increased amount of oxidative damage when fatty acids are broken down. Therefore, loading up on omega-3s may actually be adding fuel to the fire, causing additional oxidative damage in these individuals. Additional studies are needed to ascertain omega-3s' lack of benefit and possible harm in people with *E4*, but I feel the current literature provides enough support to recommend that those with one or two copies of *E4* avoid taking fish oil supplements to augment their intake of omega-3s.

Omega-3s can be good or bad, depending on your genes.

For other people, however, it appears that omega-3s have very few potential adverse reactions. And because they are most effective before Alzheimer's disease manifests and any cognitive decline occurs, I usually recommend that people who don't have the E4 variant begin taking omega-3 supplements in their early 30s if they eat fewer than three servings of fish per week.

Another complementary Alzheimer's prevention is the reduction of stress, such as by practicing yoga or meditation. Preliminary evidence shows that yoga and meditation do help protect the brain not only by correcting imbalances of neurotransmitters but by actually changing the structure of the brain itself. Studies have also found that in people with one or two copies of E4, high stress levels throughout life significantly increase the risk of cognitive decline. Although these studies found that stress led to an increased risk of cognitive decline with aging in all people, they found that having the E4 variant greatly exacerbated this effect. So it now appears that stress is another potential sledgehammer enlarging the crack in the brain's reservoir. Therefore, it is important for people with the E4 variant to be aware not only of the effects of stress on their risk of Alzheimer's but also of the various stress-reducing methods that are widely available.

MONITORING YOUR COGNITION

If a person is found to be genetically predisposed to Alzheimer's, his or her physician can administer the short Mini-Mental State Examination (MMSE) on a regular basis to monitor for any cognitive decline. The MMSE, which consists of a series of questions and mental exercises, is easily administered by any physician.

If cognitive decline is suspected, a specialized brain PET scan can provide a visual image of the brain to assess any changes consistent with pre-Alzheimer's, sometimes decades before symptoms appear. PET scans combined with genetic screening allow doctors to identify whether the early stages of Alzheimer's are occurring and how fast the disease is progressing. Based on such information, additional preventive measures can be initiated if necessary.

TREATMENT

While Alzheimer's can't be cured, the progression of the disease can be slowed temporarily with a class of medications called cholinesterase inhibitors, such as donepezil (Aricept). However, there is considerable variability in the effectiveness of this medication; approximately 1 out of 2 people show no beneficial response, and about 80 percent of this variability is thought to be due to genetics.

The *CYP2D6* gene is responsible for metabolizing many of these cholinesterase inhibitor medications, and a preliminary study has shown that variations within this gene are associated with both the effectiveness and the potential adverse effects of Aricept. For example, one variant was associated with Aricept's being effective 60 percent of the time, and a different variant was associated with it being effective only 8 percent of the time. In addition, studies currently under way are attempting to determine whether the *E4* variant also has any implications for the effectiveness of these medications. Study results on the *CYP2D6* gene are still preliminary, but I am confident that we will soon be able to genetically tailor medication treatment for Alzheimer's.

SUMMARY OF GENETICALLY TAILORED PREVENTION FOR ALZHEIMER'S DISEASE	
IF YOU DON'T HAVE ANY COPIES OF THE *E4* GENETIC VARIANT	IF YOU HAVE ONE OR TWO COPIES OF THE *E4* GENETIC VARIANT
Lifestyle	
Build a cognitive reserve	Build a cognitive reserve
Engage in physical exercise	Engage in physical exercise
Drink moderate amounts of red wine	Avoid head trauma
	Follow the Mediterranean diet
	Drink coffee or espresso
	Stop smoking cigarettes

Medications		
	Daily NSAIDs	
	Daily statins	
Complementary Prevention		
Increase intake of omega-3s from fish oil	Stress reduction techniques	
Do not take vitamin E supplements	Do not take vitamin E supplements	
Monitoring Cognition		
Routine surveillance according to the genetic risk that's determined for all cardiovascular-related diseases	Increased surveillance for and treatment of all cardiovascular diseases, including high cholesterol levels, obesity, abnormal blood pressure, and diabetes	
	Mini-Mental State Examinations	
	Brain PET scans	

We *are* fully engaged in a war with Alzheimer's disease. And this war can be won, but only if we embrace the understanding that the battle starts not when we first experience symptoms but from the day we're born. Therefore, the key to reducing your risk of Alzheimer's is to realize that this is *not* a disease of old age but one that starts causing harm to the brain much earlier in life. Prevention, therefore, requires a lifelong effort, but if we are motivated we *can* significantly decrease our risk of this disease.

We can win the war against Alzheimer's disease by fighting it throughout our life.

Example of an Alzheimer's Panel

Alzheimer's Disease Panel

- Alzheimer's disease (both sporadic and familial)
- Dementia
- Age of onset of Alzheimer's disease
- Rate of cognitive decline with Alzheimer's disease

- Effectiveness of medications used to lower risk of Alzheimer's disease
- Risk of adverse reactions to medications used to lower risk of Alzheimer's disease
- Cardiovascular disease
- Cholesterol levels
- Obesity
- Caffeine metabolism, sensitivity, and adverse reactions
- Risk of nicotine addiction and effectiveness of smoking-cessation treatments
- Risk of alcohol addiction and effectiveness of various treatments for alcohol addiction

For additional panels related to the neurological system, aging, Alzheimer's disease, and dementia, please visit www.Outsmart YourGenes.com/Panels.

10

Predict, Prevent, and Prevail over Cancer

MISCONCEPTION: Breast cancer is the only cancer that can be predicted and prevented with genetic screening.

FACT: All cancers are either partly or fully caused by variations in our genetic makeup. We can now predict whether you are predisposed to, and enact genetically tailored prevention to protect you against, *many* forms of cancer.

The premise of this book—and my work—can be summarized as follows: If we can predict, we can prevent; if we prevent, we prevail. This strategy is wholly applicable to one of the most feared and devastating diseases in all of human history: cancer.

According to a report from the World Health Organization, in 2008, 12.4 million people were diagnosed with cancer, 25 million were living with the disease, and in that year alone it had caused 7.6 million deaths. Unfortunately, in many parts of the world, cancer is a disease that's gaining strength, with the number of cases and cancer-related deaths projected to *double* by 2020. These statistics, however, do not take into account the hope that the genetic revolution now provides us—the hope that our newfound ability to predict genetic risk will enable us to prevent cancer outright and to significantly reduce the number of cancer-related deaths. The World Health Organization projects that with improved preventive strategies, one-third of all cancers can be prevented, and if cancer were detected and treated earlier, *another* third of all cancer deaths would be avoided.

The first written description of cancer appears in the world's oldest document on surgery, known as the Edwin Smith Papyrus, which was written in Egypt around 1600 BC. The papyrus describes eight cases of what was most likely breast cancer and talks about the attempted treatment with a primitive device known as "the fire drill." For any cancer that is large or appears to have spread, the papyrus simply states, "There is no treatment."

More than a thousand years later, Hippocrates, who is considered the father of medicine, also concluded that cancer was untreatable, and this belief persisted for centuries until, in the late 1700s, Scottish surgeon John Hunter put forth the radical idea that some cancers could be cured with surgery. For the first time in more than 3,000 years, the words "cancer can be cured" were finally uttered.

Since that time medicine has taken tremendous leaps forward, but cancer remains a leading cause of death for just one reason: Often it is detected too late. We now know that we can cure most types of cancer as long as we detect it early and that we can prevent many types of cancer by instituting numerous interventions. Predictive medicine takes these strategies to a new level by personalizing cancer prevention and treatment. After battling this malevolent enemy over the entire history of our civilization, we have progressed from *cancer is untreatable* to *cancer can be cured* and now to *cancer can be prevented*.

PREDICT, PREVENT, PREVAIL

When a single cell anywhere in your body turns cancerous, it suppresses or loses many of the internal safeguards that tell it to self-destruct and may also gain the ability to evade the immune system. Because of this, it is able to divide undetected and without inhibition, and eventually it grows from one cell to millions. When this occurs, the resulting mass of cancer cells can start to affect surrounding structures (for example, by pushing on them) and a few cells may break off and enter the bloodstream or lymphatic system, which enables the cancer to spread to other parts of the body in a process referred to as metastasis. These cells can then take up domicile in a new location, and once again continue their uninhibited division, increasing in number and causing significant harm. Eventually, when one or more vital organs are affected, the person dies.

Genetic variants in two types of genes are primarily responsible for causing a person to be at risk for cancer. The first of these are called proto-oncogenes, which enable a cell to divide and survive. Variants within these genes can cause the genes to become hyperactive, thereby allowing cells to divide without inhibition. The second type of genes are tumor-suppressor genes, which tell cells to slow down or even to self-destruct if they are dividing too quickly or if any of their processes get out of control. Variants in these genes can cause them to lose functionality, leaving cells free to divide with no natural inhibition.

A common analogy for explaining the functions of these two types of genes is to think of a cell as a car. The proto-oncogenes make up the gas peddle and the tumor-suppressor genes constitute the brake. Variants in proto-oncogenes cause the gas peddle to become stuck in the down position. Variants in tumor-suppressor genes disengage the brake peddle, so it doesn't work.

Doctors have only three primary methods of treating cancer once it occurs: Cut it out surgically, burn it out with radiation, or poison it with chemotherapy. Sometimes only one method is needed, and sometimes two or all three are used in combination. But whatever the method of treatment, the earlier the cancer is detected, the greater the likelihood that the treatment will be successful. Once a cancer grows too large or has metastasized, it is extremely difficult for doctors to eradicate it completely.

Predictive medicine, however, can begin to battle cancer *before* the disease even manifests. One of the first uses of predictive medicine occurred in 1995 when the University of Pennsylvania provided genetic testing for the *BRCA1* and *BRCA2* genes, both of which were shown to be associated with an increased risk of breast and ovarian cancer. At first, many people were skeptical, but now, more than 10 years later, studies have found that genetically tailored

Predictive medicine proactively fights against cancer before it manifests.

prevention based on these two genes not only reduces breast and ovarian cancer–related deaths but also prevents these cancers from developing in many people.

PROTECTING OURSELVES AND OUR FAMILY AGAINST CANCER

As a practitioner of predictive medicine I use a two-pronged approach to each type of cancer. My ultimate goal is to prevent the cancer outright by targeting individuals who are at increased risk while also having a contingency plan for detecting it at its earliest stages should it ever arise and then treating it as effectively as possible.

SKIN CANCER

Skin cancer is the most common cancer in the United States, accounting for *half of all cancers*. Fortunately, most skin cancers, even melanoma, are curable when they are detected early.

There are two types of skin cancer: melanoma and nonmelanoma (also known as basal cell and squamous cell skin cancer). While nonmelanoma skin cancer takes many years to develop and progress to the point of causing significant harm, melanoma grows quickly and has a tendency to metastasize. Once it spreads, there is close to a 90 percent chance of dying from the cancer within the next 5 years. With early

With early detection, the risk of death from melanoma can be decreased from nearly 90 percent to 1 percent.

detection, however, the risk of death from melanoma is just 1 percent. Therefore, the key to conquering melanoma is predicting who is at risk, instituting preventive measures, and making sure appropriate screening is in place throughout life in order to detect it at its earliest stages should it ever arise.

A person may be genetically predisposed to melanoma, nonmelanoma skin cancer, or both. Predictive medicine allows us to institute targeted preventive measures as early as infancy that will be tremendously beneficial throughout a person's life.

▪ GENETICALLY TAILORED PREVENTION

While most people know that changes in moles are warning signs of skin cancer, most do not know exactly what to look for. When a person is found to be genetically predisposed to skin cancer, physicians can spend time with them reviewing photographs of early-stage skin cancers so that they'll know exactly what to look for on their own bodies.

Many people also don't know that melanoma can occur on the bottom of the foot, between the toes, in nail beds, and even on the scalp. Once patients know where melanoma may occur, they can conduct proper self-examinations of their skin once a month. Because melanoma is so fast growing and deadly, I even recommend that those who are at risk ask the person who cuts their hair to let them know if they see any skin lesions or abnormalities in moles on their scalp or back of their neck.

Most people who see changes in their skin are inclined to wait a few weeks—or a few months—before seeing a dermatologist. Patients who have an increased risk for melanoma must be encouraged to develop a proactive mind-set so they see their doctor immediately after noticing any skin changes. Studies have found that people who are told by a healthcare provider that, based on genetic screening, they are at an increased risk of melanoma are very likely to take more aggressive preventive measures.

Sometimes a person's predisposition to skin cancer depends on excessive sun exposure; individuals who had one or more sunburns before the age of 18 are at highest risk. Once parents are made aware of this, they will be more likely to ensure that infants and

Personalized prevention can become a routine part of a child's daily activities.

children wear high-SPF, broad-spectrum sunscreen *every day*, even on cloudy days and all year-round. Parents should also educate their children about its importance so that wearing sunscreen becomes routine from an early age.

Another important preventive measure for people (including babies and children) predisposed specifically to melanoma is to wear sunglasses that block both UVA and UVB rays, because melanoma can also develop in the eyes.

At the time of genetic screening for melanoma, reflex analysis should also assess a child's risk of developing multiple sclerosis (MS). The conscientious use of sunscreen can reduce his or her levels of vitamin D, and low levels during childhood combined with a specific genetic variant significantly increase the child's risk of MS later in life. Physicians can advise the parents of these children to undergo yearly blood tests to assess their vitamin D levels. Reflex analysis ensures that we don't increase the risk of one disease by lowering the risk of another.

One of the most recent preventive measures is found in the form of a beverage. Recent studies have found that caffeine consumption decreases the risk of nonmelanoma skin cancer. Once it is consumed, the caffeine enters the bloodstream and eventually makes its way to the skin, where it causes cells that have been damaged by the sun to self-destruct. Drinking just one cup of tea or coffee per day can lower the risk of nonmelanoma skin cancer by as much as 30 percent, and consuming more of these beverages appears to lower the risk even further. While drinking tea or coffee is not a substitute for other preventive measures, it can be used as an adjunct by those who are at increased risk.

A GENETIC RIDDLE
What Do Red Hair, Resistance to Anesthesia, Fear of Dentists, and Skin Cancer All Have in Common?

DON'T WORRY; I'll give you the answer: They are all attributable to variants in the *MC1R* gene.

The *MC1R* gene has multiple functions, one of which is to provide pigmentation to the skin and hair. Variants in this gene are not only associated with fair skin but also with red hair and freckles. In fact, all redheads are redheads because of variants in this one gene. And because it is involved in skin pigmentation, this same gene is also responsible for whether a person will tan or just burn when exposed to sunlight.

The variants in the *MC1R* associated with having fair skin are directly linked with risk of developing both melanoma and nonmelanoma skin cancer because having fair skin increases sensitivity to sunlight. But some of the *MC1R* variants, even those that don't cause fair skin, appear to increase the amount of free radicals produced in the skin. More free radicals lead to greater DNA damage when exposed to sunlight, which means that people who have these variants will be at greater risk for skin cancer when exposed to sunlight regardless of skin tone.

The *MC1R* gene is involved in a number of other interesting traits as well. The same variants that cause people to have red hair also cause them to require higher doses of both general (puts you to sleep) and local (numbs you to pain) anesthesia. Dr. Matthew Giudice, an anesthesiologist in San Francisco, notes that these individuals can require as much as 20 percent more anesthesia to achieve the desired effect.

The *MC1R* gene is not only very active in the skin but also plays a role in the brain, where it's involved in the processing of

continued

pain, fear, and anxiety. Therefore, variants in this gene are responsible for dictating a person's response to anesthesia and for the way their brain handles pain. This is certainly one busy gene!

If you felt pain every time you visited your dentist, despite the fact that he or she was assuring you that your mouth should be numb, it wouldn't be long before you began to associate dentistry with pain. With 15 percent of all people experiencing *extreme* fear and anxiety, and 45 percent experiencing moderate fear when they go to the dentist (or even think about going to the dentist), fear of dental care is a very real thing. Getting to the bottom of why people have this fear is important because studies have shown that individuals who avoid the dentist have considerably more dental problems in the long run.

According to a 2009 study, the same genetic variants that make a person more resistant to local anesthesia (such as Novocain) cause them to have such increased pain during dental procedures that they actually avoid dentists much more often than people without these variants.

Variants within the *MC1R* gene do not, however, appear to cause resistance to the pain-reducing effects of other medications, such as codeine and Vicodin. Therefore, premedicating with one of these and concurrently using higher doses of a local anesthetic may help provide significant pain relief during dental procedures, thereby limiting, and potentially curing, the anxiety and fear associated with dental care.

COLORECTAL CANCER

Colorectal cancer is currently the third most prevalent type of cancer. The lifetime risk of colorectal cancer for people in the United States is approximately 5 percent, and more than 75 percent of all cases occur in people without known risk factors, such as a family history of the disease. Like skin cancer, however, *no one* should ever die from colorectal cancer because we have the ability to detect and remove it successfully as long as we get to it in its early stages.

Even though the general guidelines state that all people over the age of 50 should have a colonoscopy every 10 years, only 50 percent of people follow this recommendation.

As long as colorectal cancer is detected in its early stages, it can almost always be cured.

As a result, only 40 percent of colorectal cancers are diagnosed at an early stage. Unfortunately, 35 percent of people are diagnosed at an intermediate stage, which equates to nearly a 1 in 3 chance of dying within 5 years, and almost all others are diagnosed at an advanced stage, which is associated with a 90 percent chance of dying within 5 years.

Colorectal cancer can occur because of a rare genetic disease, such as familial adenomatous polyposis or Lynch's syndrome or because a person is genetically predisposed to this type of cancer and has been repeatedly exposed to certain nongenetic factors (specific foods) that cause the cancer to form.

Discovering that you are at increased risk of colorectal cancer is incredibly empowering for several reasons. First, there are numerous preventive measures that enable you to lower your risk. Second, there are several types of screening exams that can detect precancerous and cancerous changes in your colon so that they can be removed before they become life threatening. Third, knowing you're at risk will motivate you never to skip or put off a colonoscopy. And fourth, it motivates your doctor to be more proactive in alerting you to the need for screening exams and making sure you follow through.

■ GENETICALLY TAILORED PREVENTION

Many lifestyle and dietary changes have been shown to reduce the risk of colorectal cancer. One of the most widely known is to decrease consumption of red meat and processed meats, both of which have been shown to increase the risk of colorectal cancer when either is consumed on a regular basis.

Red meat is thought to cause colorectal cancer because cooking it at high temperatures creates a number of potentially harmful substances that are ingested when the meat is consumed. Studies have found that variants in two genes, *NAT1* and *NAT2*, substantially increases a person's risk of colorectal cancer if they eat more than one serving of red meat per day. This occurs because the variants cause those genes to convert more of red meat's *potentially* harmful substances into *actual* harmful substances, which can then damage the cell. Because these two genes are very active in the cells lining the colon, this is where the damage occurs, greatly increasing the likelihood of cancer forming in the damaged cells. Thus people with these variants should avoid eating red meat on a regular basis.

Physical exercise can also reduce a person's risk of colorectal cancer by as much as 50 percent, making it one of the most powerful preventive measures anyone can institute. Even brisk walking done a few times a week has been shown to be effective, and the more you exercise, the more you reduce your risk. And, best of all, even if you've lived a sedentary life up to now, you can still decrease your risk of colorectal cancer if you start exercising on a regular basis.

That said, however, being overweight or obese significantly increases a person's risk of colorectal cancer independent of whether the person engages in regular exercise. While it may seem counterintuitive, many people who are overweight or obese are physically active. But studies have shown that being overweight or obese trumps exercise in terms of colorectal cancer risk. Therefore, it is crucial for people predisposed to colorectal cancer to lose those extra pounds— or to avoid gaining them in the first place. There also seems to be a connection between increased fat around the abdomen area and an increased risk of this cancer.

In terms of medication, aspirin has been shown to decrease the risk of colorectal cancer by directly limiting the growth of cancerous cells. But because daily aspirin use can cause stomach ulcers and other complications, this preventive measure, as with all medications, should be undertaken only in consultation with a physician.

Other useful strategies for reducing one's risk are to be sure that vitamin D levels are normal and to eat more garlic. People with normal levels of vitamin D have a lower risk of colorectal cancer than those whose levels are below normal, and people who eat the equivalent of one or more whole cloves of garlic each day also appear to have a much lower risk of the disease. Current research also indicates that freshly chopped, cut, or crushed garlic appears to provide greater anticancer benefits than garlic supplement pills.

When it comes to screening exams, the most common and most effective is a colonoscopy, which involves inserting a very thin snakelike scope into the rectum and carefully maneuvering it all the way to the end of the colon so that the physician can clearly view the inside and take a biopsy of anything that looks abnormal. A colonoscopy is a powerful weapon in our battle against colorectal cancer and the number of deaths from this disease would be drastically reduced if everybody had one as indicated. Even though the procedure may mean a few hours of discomfort every 5 to 10 years, colorectal cancer can end your life prematurely—so spending a few *hours* now to gain *decades* later is definitely not a bad trade-off.

There are also newer techniques, such as virtual colonoscopies, which use a CAT scan or an MRI to look inside a person's colon and screen for signs of cancer. I do not recommend virtual colonoscopies for patients who are predisposed to colon cancer because if a polyp or other type of abnormality is identified, a traditional colonoscopy is still required for removal or biopsy, thereby putting the patient through two procedures instead of one.

Finally, you may have heard about a procedure called a flexible sigmoidoscopy. This is *not* appropriate for people genetically predisposed to colorectal cancer because it views only the lower third of the colon. While people may choose flexible sigmoidoscopy because it takes less time and is less complicated, it does not provide full screening coverage.

For people predisposed to colorectal cancer, the American Cancer Society explicitly recommends that only traditional colonoscopy be used. I usually suggest that people who are found to be at increased risk have a traditional colonoscopy every 5 years starting at the age of 40. However, if a person has a family history of colorectal cancer, the first colonoscopy should be done 10 years before the age at which the family member was diagnosed and every 5 years thereafter. For some rare causes of cancer, colonoscopies starting in the preteen years are recommended because the cancer occurs much earlier in these individuals.

Predictive medicine personalizes cancer screening recommendations.

There is also a technique called a fecal occult blood test (FOBT), which tests for hidden blood in the stool. Because blood indicates the possibility of a cancerous growth in the colon, a positive test indicates that the person needs to have a follow-up colonoscopy. For people predisposed to colorectal cancer, I usually recommend an FOBT each year between colonoscopies just to be sure that nothing was missed during their previous colonoscopy.

PROSTATE CANCER

Prostate cancer is an extremely prevalent disease, with early stages detectable in more than 30 percent of men in their 50s and 70 percent of men in their 80s. However, because most prostate cancers grow very slowly, only about 3 percent of men actually die of the disease. Nonetheless, because of the cancer's proximity to an important organ, and because the treatment is sometimes associated with side effects (including incontinence and erectile dysfunction) that can significantly alter a man's quality of life, prostate cancer remains a feared disease.

A number of genetic variants have been found to increase the risk of prostate cancer. Some studies have investigated the usefulness of analyzing all known prostate cancer–associated variants at once and have found that combining the analysis of multiple variants with a person's family history of prostate cancer has very strong predictive power. For example, the general lifetime risk of a 55-year-old man is 13 percent. But by analyzing a group of genetic variants and family

Combining family history with genetic screening allows predictive medicine to provide a more accurate assessment of prostate cancer risk.

history, we are now able to determine that the actual lifetime risk for an individual with very few harmful variants and no family history of the disease can be as low as 6 percent and that the risk for a man who has many harmful variants and a family history of the disease can be as high as 41 percent.

Other studies have also found associations between variants and specific types of prostate cancer. As more research is completed, we'll be able to start predicting the aggressiveness of prostate cancer should it ever arise. This is incredibly important information because men who have very slow-growing prostate cancer sometimes don't need to do anything at all while those who have the more aggressive, life-threatening form do need to undergo treatment.

▪ GENETICALLY TAILORED PREVENTION

The drug finasteride (Proscar) has been shown to lower the risk of developing prostate cancer by 25 percent. The medication works by reducing the amount of testosterone in a man's body, and since prostate cancer feeds off testosterone, reducing the hormone level also reduces the cancer risk. In essence, finasteride deprives prostate cancer of its food so that it can't grow. Finasteride is an effective preventive option that should be considered by men who are at considerably increased risk of prostate cancer.

As with colorectal cancer, there are also a number of lifestyle modifications that have been shown to decrease risk. One such prevention is to increase consumption of foods that contain lycopene, which is primarily found in tomatoes and products that contain tomatoes, such as tomato sauce. Eating two or more servings of these foods every week has been associated with a 35 percent reduction in risk for prostate cancer and an even more substantial reduction in the risk of aggressive prostate cancer.

Another preventive strategy that has been widely studied is increasing one's consumption of omega-3 fatty acids, which are found in fish oil. We've already discussed the fact that omega-3s

Preventive measures can lower a man's risk of prostate cancer.

significantly decrease the risk of heart disease, but they also appear to decrease the risk of prostate cancer, especially aggressive prostate cancer.

Preliminary studies have also found that a specific genetic variant in the COX2 gene is associated with the degree to which omega-3s may reduce a man's risk of prostate cancer. Individuals who have this variant were found to have a substantially increased risk of prostate cancer if they consumed low levels of omega-3s. However, if people with the same variant consumed high levels of omega-3s, such as by eating fatty fish on a regular basis, they were actually protected against prostate cancer and their risk was lower than that of the general population. If these results are confirmed by further studies we will have another highly effective preventive measure against prostate cancer for those people with this particular variant.

The two routine screening methods currently available for prostate cancer is a blood test for prostate-specific antigen (PSA) and also the infamous digital rectal exam (DRE). PSA levels increase with prostate cancer; if the level is abnormally high, an ultrasound of the prostate is performed. If the physician sees a mass on the prostate, he or she will then take a biopsy. PSA levels, however, are affected by a number of other harmless conditions, which significantly reduces the accuracy of this test for the detection of prostate cancer.

The DRE, on the other hand, is a way for the physician to physically feel the prostate in order to check for any abnormal enlargement. If an abnormality is detected, then a biopsy is performed.

My screening recommendation for people at increased risk of prostate cancer is to start having an annual DRE 10 years earlier than normal (starting at the age of 40). However, in terms of the PSA, I recommend to do nothing different from people who do not have an increased risk—that is, to discuss with your doctor having an annual PSA blood test starting at age 50. Because a PSA test is not accurate enough to be used earlier or more often, men with an increased risk for prostate cancer should focus primarily on preventive strategies, such as taking finasteride, eating more tomatoes, and being adamant about having an annual medical checkup that includes a DRE.

PROTECTING THE BREATH OF LIFE AGAINST ADDICTION AND LUNG CANCER

Lung cancer is the number one cause of cancer-related deaths, and cigarette smoking is directly responsible for about 85 percent of all cases of lung cancer. Therefore, freeing the world of its nicotine addiction is the most effective weapon we have in fighting lung cancer.

Genetics explains why some people who smoke will eventually die from lung cancer, others who smoke will live cancer free, and some nonsmokers will also die of lung cancer. While researchers are still working to determine the exact genetic causes of lung cancer in people who never smoke, we've already made significant headway

Because approximately 85 percent of cases of lung cancer are attributable to cigarette smoking, conquering nicotine addiction is the most effective preventive weapon we have in fighting lung cancer.

in terms of understanding the genetic association among nicotine addiction, smoking, and lung cancer.

Because smoking is directly associated not only with lung cancer but also with a myriad of other life-threatening diseases, it is the number one cause of preventable death in the United States. Cigarettes are responsible for 1 in 5 deaths in the United States each year, and in the last century, smoking has killed approximately *100 million* people worldwide. Smoking-related illnesses and death result in a tremendous cost to our society, with a total of almost $200 billion *per year* attributable to healthcare costs and lost productivity in the United States alone.

Nicotine is *extremely* addictive, especially for people with a specific genetic predisposition. Almost 20 million smokers attempt to quit each year, but approximately 80 percent relapse within the first month and 97 percent within the first 6 months. So many people are addicted that, throughout the world, 10 million cigarettes are sold *every single minute.*

The majority of our risk for nicotine addiction is determined by our genes, and we can now start to use knowledge of our genetic code not only to predict risk of nicotine addiction but also to create genetically tailored strategies for quitting. To defeat lung cancer, we have to not only understand the cancer but also the addiction. Understanding the fundamental genetic reasons for addiction empowers us to outsmart this adversary through genetically tailored prevention and treatment.

■ GENETICALLY TAILORED PREVENTION

The predictive medicine approach I've developed for dealing with lung cancer can be divided into three distinct strategies—one applies to current smokers, another applies to people who are genetically predisposed to nicotine addiction but who are currently nonsmokers (such as children and former smokers), and the third applies to nonsmokers who are exposed to secondhand smoke.

Unfortunately, we need more research before we can begin to predict and prevent lung cancer in nonsmokers. And even after we discover a number of variants associated with an increased risk of the disease in those who don't smoke and who are not exposed to secondhand smoke, that information won't be actionable until we come up with a way to modify the risk, such as with screening exams, taking a specific medication or vitamin, or by changing a lifestyle factor such as diet or exercise.

There are currently no screening exams for lung cancer that have been shown to save lives and no specific nongenetic factors other than not smoking and avoiding exposure to asbestos and radon that have been shown to reduce a person's risk of lung cancer. Even substances once thought to decrease risk, such as multivitamins, vitamin C, vitamin E, and folate, have now been found to have no effect at all.

STRATEGY 1: FOR CURRENT SMOKERS

Smoking is bad for everyone, but some people are more susceptible than others to developing cancer. Some people have genetic variants that either lead to higher levels of carcinogens (a carcinogen is anything associated with increasing a person's risk of cancer) over longer periods of time or that decrease their DNA's ability to repair itself so that the harm done by carcinogens accumulates more significantly over time. Being able to tell smokers that they are at a substantially higher risk of cancer because they smoke has the potential of making them truly want to quit rather than simply restating the widely known blanket statement that smoking is bad for everyone.

Because being able to quit smoking partly depends on a person's genetic code, genetically tailored smoking cessation programs have the potential to significantly increase the chances that a smoker will be able to quit. For example, one of the most common FDA-approved medications prescribed for smoking cessation is bupropion (Zyban and Wellbutrin), which decreases the urge to smoke and increases the likelihood that a person will be able to quit. Nevertheless, approximately 60 percent of people who use bupropion eventually relapse, indicating that this medication, like most, is effective only for a subset of people.

Studies have found that genetic variants in the *CYP2B6* and *DRD2* genes are associated with differences in the effectiveness of bupropion for smoking cessation. Approximately 50 percent of African Americans, 45 percent of Caucasians, and 25 percent of Asians have a specific variant within the *CYP2B6* gene that makes the gene less active. Bupropion treatment is much more likely to be effective for these people because *CYP2B6* is responsible for breaking down bupropion, and when the gene is less active, the bupropion sticks around longer in the body, exerting more of a beneficial effect. Without that variant, the bupropion is processed more quickly, substantially decreasing its overall effectiveness.

It is interesting that some of the same studies have shown that in people for whom bupropion is likely to be ineffective, using nicotine-replacement therapy in the form of gum, a patch, or nasal spray can be an effective way to quit. And other variants even determine which type of nicotine-replacement therapy may work best.

The choice of therapy used to help someone quit smoking can be tailored to that person's genetic makeup.

For example, a preliminary study has found a variant that increases a person's chances of quitting with the use of nasal spray rather than the patch. The most likely reason is that this variant is associated with a person's experiencing a greater rush from nicotine when he or she smokes cigarettes. Because nicotine from a nasal spray enters the blood more quickly than that from either the patch or the gum, it more closely mimics the rush the person with this variant is used to. Other studies, however, have found that a number of variants in other genes are associated with a greater chance of quitting if a person uses the patch. The patch delivers higher levels of nicotine than either gum or nasal spray, and for people with these variants, higher levels of nicotine equate with improvement in mood and less weight gain, thereby increasing the likelihood that this specific therapy will work for them.

IF ONLY POPEYE HAD PREDICTIVE MEDICINE

POPEYE HAD three true loves: Olive Oyl, spinach, and smoking his pipe. While his copious consumption of spinach protected him from Brutus's attacks, it, unfortunately, didn't protect him from a much more deadly adversary: lung cancer.

Like Popeye, many smokers are unable to quit, but predictive medicine can still help them institute genetically tailored strategies that may decrease their risk of lung cancer even while they continue smoking. For example, 20 to 50 percent of all people have variants in their *GSTM1* and *GSTT1* genes. If a smoker has variants in these genes and also eats one or more servings of cruciferous vegetables—such as broccoli, cabbage, Brussels sprouts, or collard greens—at least once a week, they'll have a decreased risk of lung cancer compared to smokers who either don't have these variants or don't eat cruciferous vegetables on a regular basis.

These genes are responsible for metabolizing the isothiocyanates found in these vegetables. Isothiocyanates protect the body against cancer, and the more efficiently they are metabolized, the less likely they are to accumulate in the body. Therefore, when variants in these genes prevent them from functioning, the beneficial isothiocyanates build up to more effective levels. It appears, however, that the isothiocyanates exert their anticancer effects only when they are ingested at about the same time the body is being exposed to a carcinogen, such as if the person smokes within a few hours before or after the meal.

Studies have found that smokers in general tend to eat fewer vegetables than nonsmokers, and, like Popeye, they sometimes eat the wrong vegetables. Spinach is a *non*-cruciferous vegetable and does not contain isothiocyanates, which means that even for smokers with *GSTM1* and *GSTT1* variants, it is not effective for fighting lung cancer. If he had these variants, predictive medicine could have guided Popeye to eat something more beneficial.

STRATEGY 2: FOR CURRENT NONSMOKERS WITH A PREDISPOSITION TO NICOTINE ADDICTION

While it might seem logical to assume that nongenetic factors are primarily responsible for whether a person smokes, it turns out that our genes determine 60 percent of why we become addicted to smoking. This is why some people can smoke on and off for a few years and then just quit, whereas others become hooked for life after their first pack.

Recently, variants in the *CHRNA3* and *CHRNA5* genes were found to increase one's risk of becoming addicted to nicotine. When a person smokes, the nicotine from the cigarette is inhaled into the lungs, enters the bloodstream, and travels to various parts of the body, including the brain. These two genes are responsible for producing receptors in the brain that bind to nicotine, and variants within these genes cause the brain to be much more susceptible to nicotine addiction. A preliminary study found that the variants are associated with experiencing a euphoric buzz when a person smokes a cigarette for the first time and that this effect correlates with addiction and the long-term continuation of smoking.

Researchers have also started to find that there are genetic variants associated with the *age* at which a teenager or young adult is most likely to start smoking and with the number of cigarettes a person is likely to smoke each day. Environmental factors, such as peer pressure and experimentation, were once thought to be the only reasons people start smoking, but it now appears that the underlying cause also includes a genetic factor.

As more research is conducted, we'll be able to predict not only who is at significantly increased risk for becoming addicted to nicotine but also which adolescents are most likely to experiment and at what age they are most likely to begin smoking. Once we are able to predict these traits, we can institute targeted prevention, such as specialized education designed to fully inform children about their *personal* risk. And reflex analysis will provide information not only about risk of nicotine addiction but also about risk for all the diseases associated with nicotine addiction, such as cancer. The potential effectiveness of this type of prevention relates to its being comprehensive, integrated, and personalized.

STRATEGY 3: FOR NONSMOKERS EXPOSED TO SECONDHAND SMOKE

In North America and many European countries, exposure to secondhand smoke has been significantly reduced by bans on smoking in public. However, many spouses and children of smokers are still exposed to secondhand smoke, as are those who work in places like casinos and bars where smoking is sometimes still allowed. And public exposure to secondhand smoke still remains a health hazard in many nations throughout Asia, Southeast Asia, and the Middle East.

Preliminary studies have identified a variant in the GSTM1 gene that increases the risk of lung cancer in people exposed to secondhand smoke by more than 100 percent. To bring the significance of this statistic into perspective, exposure to secondhand smoke raises the risk of lung cancer in the general population by only about 20 percent. This means that compared to most people, those with this variant in their GSTM1 gene appear to be at much higher risk of lung cancer due to exposure to secondhand smoke.

The GSTM1 gene plays a pivotal role in detoxifying carcinogens, and the variant prevents the gene from functioning, thereby leading to a potential buildup of harmful substances in the body. If these findings are verified, physicians will be able to identify nonsmokers who are at substantially increased risk of lung cancer from exposure to secondhand smoke and to counsel them and their families on the need to avoid cigarette smoke. When this variant occurs in conjunction in a person who also has variants that react well to cruciferous vegetables, the patient could also make dietary changes.

While additional research is needed to verify these findings, the initial studies offer hope that one day soon we'll be able to predict the true effects of exposure to secondhand smoke on nonsmokers.

■ GENETICALLY TAILORED TREATMENT

Genetic analysis is becoming increasingly capable of providing guidance for individually tailored treatment of lung cancer if the disease should ever occur. Laboratory studies have already found that when people with specific genetic variants were given a particular type of chemotherapy, their probability of survival significantly increased. As more research is conducted, this kind of information will move from the laboratory to the hospital so that we can use the full power of a person's genetic code to most effectively treat lung cancer.

GOING ON THE OFFENSIVE IN OUR BATTLE
AGAINST BREAST AND OVARIAN CANCERS

The first screening test for any type of cancer was developed for cervical cancer in 1923 by George Papanicolaou (for whom the Pap smear is named). However, the medical establishment was highly skeptical of this new screening test, and, therefore, it wasn't widely used for almost 40 more years, until the American Cancer Society started to promote it in the 1960s. Since that time, this one screening test has reduced the rate of death from cervical cancer in the United States by about 70 percent, showing just how powerful screening tests can be in helping battle disease.

Many other screening tests have been developed for cancers that affect women. Cancer doesn't just appear overnight and there is usually a latent phase that involves changes to cells over many months or even years. The cells usually go through a precancerous stage during which time they are on their way to becoming cancer but haven't yet become malignant. Over time, some of these cells will be cancerous. When that occurs, the rate at which the cells multiply increases dramatically, and the cancer starts to grow. While early-stage cancer is confined to a small and well-defined area of the body, more advanced, late-stage cancer is larger and either has or is about to spread. The outcome of surgical treatment of most cancers, including breast and ovarian cancer, is excellent if the disease is detected when it is precancerous or during its early stages. As the cancer grows and becomes more advanced, surgery, medications, and radiation become less effective, and the overall prognosis worsens. The goal of nongenetic screening tests, therefore, is to identify precancerous changes or the very early stages of actual cancer so that appropriate measures can be implemented to stop the disease from progressing.

Breast cancer is the second most common type of cancer (after lung cancer) and is responsible for almost 1 percent of deaths from all causes worldwide. Without considering genetics or other risk factors, the general lifetime risk of breast cancer for women in the United States is about 12 percent, meaning that approximately *1 in 8* women will be diagnosed with the disease during their lifetime. And 1 out of every 35 women in the United States will die from this disease. If a woman has a genetic variant in a gene related to breast cancer, then her risk of breast cancer can be increased by more than 500 percent, which means that her actual lifetime risk can be as high as 65 to 80 percent.

One statistic that many people are unaware of is that breast cancer is actually the number one cause of death due to cancer in women between the ages of 15 and 29. And unlike cancer in women older than 50, the incidence of breast cancer in this younger age group is increasing. Very often, these younger women have one or more genetic variants that predispose them to the disease.

The same genetic variants that predispose a woman to breast cancer also increase her risk of ovarian cancer, and many of the preventive measures are similar for both. A woman's general lifetime risk of ovarian cancer is 1.4 percent, but a single genetic variant can increase that risk to as much as 40 percent.

As with other cancer studies, research into breast cancer has shown that learning one is genetically at risk for breast cancer does not increase long-term anxiety. In fact, most studies show a decreased level of anxiety after testing. One small preliminary study has even found that children, teenagers, and adults under the age of 25 did not suffer any negative psychological consequences when told that their mother had a variant that significantly increased her risk of breast cancer and that, therefore, they too might be at increased risk. All the study subjects conveyed an appropriate level of understanding of the information and many of them also instituted preventive lifestyle changes for themselves.

After years of research, our war against breast cancer has started to show glimmers of success. For women older than 29, the rate of new cases is declining by about 3.5 percent annually, and the rate of death from breast cancer is declining by 2 percent each year. These statistics indicate that prevention and early detection are working, that treatments are improving, and that we can certainly do better. While doctors have started to chip away at breast cancer, we must now make a concerted effort to target this disease with all of our technological might to decrease the rates of new cases and deaths even more significantly. Predictive medicine provides us with this capability.

UNDERSTANDING THE GENETICS OF BREAST AND OVARIAN CANCER

BRCA1 and *BRCA2* (collectively referred to as *BRCA1/2*) are undoubtedly the two most infamous genes in the world. In North America approximately 1 in every 400 people has a *BRCA1/2* genetic variant. Variants in these two genes are responsible for about 5 percent of all breast cancers and 10 percent of all ovarian cancers. And while they are responsible for only a fraction of all breast and ovarian cancers, when they do occur, they increase risk of these diseases quite dramatically.

The association between variants in these genes and an increased risk for breast and ovarian cancers became apparent in the mid-1990s, and almost immediately, physicians started to use this new weapon in the war against breast and ovarian cancer.

However, the *BRCA1/2* genes have gotten a bad rap because they are *not* intrinsically evil. In fact, when they are functioning normally they protect our bodies against the damage that occurs in specific types of breast and ovarian cells. The *BRCA1/2* genes are known as tumor-*suppressor* genes because they *repair* DNA if it becomes damaged. And because environmental factors (such as radiation and toxins) are constantly damaging our DNA, having a DNA repair mechanism is extremely important.

Their evilness manifests only when a variant in one of these genes prevents it from functioning properly so that DNA is not repaired and, over time, DNA damage may accumulate. Eventually, the DNA in a single cell may become damaged in such a way that it starts sending abnormal signals to the cell to multiply nonstop, and this causes the cell to become cancerous.

Because it takes many years for DNA damage to occur and accumulate, cancer in people with *BRCA1/2* variants doesn't manifest for a few decades. Therefore, these individuals likely have already had children and potentially passed on the cancer-associated variant.

The following table summarizes the probability of having a cancer-associated variant in either the *BRCA1* or *BRCA2* gene, based on a variety of factors.

PROBABILITY OF HAVING A CANCER-ASSOCIATED VARIANT IN *BRCA1/2*	
General female population	0.25%
Family History	
Breast cancer in one first- or second-degree relative	6.25%
Breast cancer in two first- or second-degree relatives	12.5%
Breast cancer in more than two first- or second-degree relatives	50%
Breast and ovarian cancers in the same family	40%
Personal History for Females	
Breast cancer after age 40	2%
Breast cancer before age 40	10%
Breast cancer before age 30	23%
Ovarian cancer at any age	10%
Female Ethnicity	
Asian American who has had breast cancer before age 65	0.5%
African American who has had breast cancer before age 65	2.6%
Hispanic American who has had breast cancer before age 65	3.5%
Ashkenazi Jews	2.5%
Ashkenazi Jew who has had breast cancer after age 40	10%
Ashkenazi Jew who has had breast cancer before age 40	33.3%
Ashkenazi Jew who has had ovarian cancer at any age	33.3%
Males	
Man with breast cancer at any age	5%
Ashkenazi Jew with breast cancer at any age	20%

USING COMPREHENSIVE SCREENING TO GAIN AN INVALUABLE ADVANTAGE

Historically, physicians have frequently pondered, "Who will benefit from screening?" Based on the cost-benefit ratio, screening couldn't be offered to everyone and instead was primarily recommended only for individuals whose risk, based on family history, was determined to be greater than a certain percentage. Now, because comprehensive genetic screening is so much more cost effective and because we are able to screen for an almost unlimited number of variants, we can include breast and ovarian cancer risk screening in a number of panels, such as the newborn panel, children's panel, women's health panel, men's health panel, and the cancer panel. Instead of trying to determine who might benefit most, we can screen anyone interested, and let the person's genetic code speak for itself. And even if the person is not found to be at increased risk for these two cancers, the panel will still provide valuable information about other diseases.

There are many genes besides BRCA1/2 involved in determining breast cancer risk. A recent study found that combining family history with an analysis of variants in more than 15 different genes was significantly more accurate than other methods currently being used for assessing risk.

While some of these variants increase risk independent of BRCA1/2, others interact with these two genes to increase their risk (called modifier genes). Therefore, an accurate

While genetic variants within the *BRCA1/2* genes are the ones most infamously associated with breast cancer, there are actually many different genetic variants in a large number of other genes that are also associated with an increased risk of breast cancer, and an accurate analysis of risk should take them all into consideration.

risk assessment is predicated on analyzing not only BRCA1/2 variants but also the modifier genes.

For example, a variant in the RAD51 gene substantially increases the risk of breast cancer only in people who also have BRCA2 variants. The RAD51, a DNA repair gene, augments the disruption of DNA repair caused by BRCA2. A combination of variants in both of these genes increases a woman's lifetime risk of breast cancer from just under 50 percent to greater than 80 percent.

Variants in the CHEK2 gene work independently to increase the risk of breast cancer by more than 25 percent. The gene acts like a gatekeeper by regulating how rapidly a cell can divide if it is ever exposed to radiation or other factors that cause damage to the cell's DNA. When DNA damage accumulates but is not properly repaired, this gene causes the cell to self-destruct. Variants in CHEK2, however, render the cell's self-destruct system less effective. If radiation or other damaging factors occur and a cell starts to multiply out of control, if the CHEK2 gene contains a variant that causes it to malfunction, the gene won't function properly and the abnormal cell may not self-destruct, allowing the cell to potentially turn cancerous.

The ATM gene has a similar function, and approximately 1 in every 10 women with breast cancer will have a harmful variant in her BRCA1/2, CHEK2, or ATM gene.

Because the American Cancer Society recommends increased surveillance and prevention in women who have a greater than 20 percent lifetime risk of breast cancer, having one or more variants in any one of these genes can indicate the need for genetically tailored prevention.

THE DISEASE MATRIX AND REFLEX ANALYSIS IN ACTION
Alcohol, Genes, Heart Disease, and Breast Cancer

AT THIS point it is clear that we can now gather a tremendous amount of information by analyzing our genes, but this new-

continued

found capability also results in an almost overwhelming amount of data. Discussing alcohol's effects on heart disease and breast cancer provides a clear example of why we need to use analytical technologies that are able to integrate multiple pieces of information to arrive at straightforward, actionable results.

If moderate alcohol intake protects us against heart disease, is it also useful in protecting us against cancer? The overwhelming answer is no; not only is it not protective but alcohol actually *increases* our risk of different types of cancer, including breast cancer.

Alcohol increases the risk of breast cancer most likely because it causes estrogen levels to increase, and increased estrogen levels are directly correlated with an increased risk of breast cancer. Another possible reason is that alcohol intake can lead to decreased folate levels, and low folate levels negatively affect DNA's ability to function properly. Therefore, it now appears that some of the increased risk of breast cancer with alcohol intake can actually be mitigated by consuming at least 300 micrograms of folate per day from a multivitamin or by consuming folate-rich fruits and vegetables.

Studies have shown that for the general population, drinking a single glass of alcohol per day increases the risk of breast cancer by about 10 percent, and for those who drink between two and five glasses per day the risk increases by 40 percent. While the increased risk associated with just one glass of alcohol may not appear to be substantial, it now appears that for some women alcohol intake may do much more harm.

Preliminary studies are now showing that women who have specific variants in both their *GSTM1* and *GSTT1* genes and who consume one or more alcoholic drinks per day may increase their risk of breast cancer by more than 100 percent. These genes are important for detoxifying the body of harmful substances like those produced when the body metabolizes alcohol. When these variants occur, the genes don't function at all, so the toxins may build up.

Clearly then, tailoring prevention to a person's genetic information can actually get quite complex. If, for example, a person at increased risk of heart disease is found to have variants in her *GSTM1* and *GSTT1* genes, thereby significantly increasing her risk of breast cancer with alcohol intake, alcohol will not be recommended. But if her risk of heart disease is significant, and she doesn't have these particular variants, she will be advised that having one alcoholic drink a day, preferably red wine, along with adequate folate intake, is likely to be beneficial, although drinking more could be detrimental. By taking into consideration a multitude of diseases, genes, and nongenetic factors, reflex analysis using the disease matrix allows for the most effective, genetically tailored preventive strategies to be developed and implemented.

GENETICALLY TAILORED PREVENTION

I believe we have to stop thinking of breast cancer as a disease only women need to worry about and understand that with predictive medicine the battle against this disease is *no longer* gender specific.

Both women and men can be empowered by genetic screening for their risk of breast cancer. While I am at considerably lower risk of breast cancer because I'm a man, discovering I had a breast cancer variant would not only alert me to institute preventive measures in my own life but would also allow me to understand the risk to my future offspring. If I have a breast cancer variant, even if breast cancer never affects me, it could very well affect any

With predictive medicine and comprehensive genetic screening, both women and men are able to fight side by side in the battle against breast cancer.

daughters I might have in the future. Knowing this allows me to decide whether I want to institute preventive measures, such as using preimplantation genetic screening to prevent passing on the variant. Perhaps I'd instead choose to analyze my daughter's genes at birth and, if she is found to have inherited the harmful variant, implement preventive measures throughout her entire life.

If you are found to be at increased risk of breast or ovarian cancer, there are highly actionable preventive measures that have been shown to significantly reduce the risk of these diseases. Because breast cancer is considerably more prevalent than ovarian cancer, it has been the primary focus of many of the studies conducted to date. However, whenever there is scientific data regarding the prevention of ovarian cancer, it will be included in the recommendations that follow.

■ Lifestyle

No lifestyle factor has been shown to affect a person's risk of breast cancer more than exercise. Engaging in cardiovascular exercise such as swimming, biking, playing tennis, jogging, or even brisk walking 4 or more hours per week lowers the risk of breast cancer by about 30 percent.

One reason for this is that women who exercise have lower levels of estrogen than those who are sedentary. As the medical community has learned over the past few decades, higher estrogen levels increase a person's risk of breast cancer, most likely because estrogen can directly stimulate cells to proliferate, and encouraging cells to divide can promote the development of cancer. In fact, it now appears that *anything* that increases estrogen levels also increases the risk of breast cancer.

Cardiovascular exercise can significantly decrease your risk of breast cancer.

Exercise on its own has been shown to reduce the risk of breast cancer, and not becoming overweight or obese after menopause is another way to reduce one's risk later in life. When the percentage of body fat increases, so does a person's estrogen levels, because fat cells convert other hormones into estrogen. This association, however, is observed only after menopause, because before menopause the amount of estrogen in a woman's body primarily depends on the high levels produced by her ovaries, and fluctuations related to levels of body fat play a comparatively smaller role.

GENETICALLY TAILORED TEA TIME

OTHER THAN water, tea is the most widely consumed beverage throughout the world. This is a good thing because it appears that tea has some rather amazing health benefits—at least for some people. Similar to red wine, tea is an excellent source of polyphenols, which protect not only against cardiovascular disease but also against cancer. After red wine, the highest overall concentration of polyphenols are found in matcha (powdered green tea leaves), green tea, oolong, and black tea, in that order.

Studies have shown that the polyphenols found in both wine and tea protect against cancer, most likely because they are excellent natural antioxidants that defend the body against free-radical damage that would otherwise put cells at greater risk of turning cancerous. In addition, tea-specific polyphenols appear to stop the proliferation of the extra blood vessels that a cancer requires to support its increased nutrient needs. Because of these effects, studies have shown that drinking three or more cups of tea per day lowers the risk of stomach, prostate, oral, esophageal, bladder, and breast cancer by as much as 50 percent.

It is interesting, however, that one's individual genetic makeup may dictate whether he or she is able to reap health benefits from tea. A variant in the *COMT* gene, which is involved in the processing of tea-specific polyphenols, causes the gene to become less active, meaning the polyphenols aren't processed as rapidly so they have a greater effect on the body.

continued

In 2003, researchers at the University of Southern California Keck School of Medicine conducted a preliminary study that investigated the association between tea consumption, breast cancer, and variants in the *COMT* gene in Japanese American women. They found that women who were tea drinkers and who had the low-activity *COMT* variant had a 50 percent lower risk of breast cancer but that drinking tea had no discernible effect on the risk of breast cancer in women who didn't have the variant.

In a follow-up study in 2005, another research team at the same institution investigated whether variants in the *ACE* gene were also associated with tea drinking and a reduced risk of breast cancer. The *ACE* gene was chosen because it contains one variant that causes the production of a higher amount of free radicals and another variant that significantly reduces the production of free radicals. It's been known that women with the *ACE* variant that is associated with lower free radical production have a lower risk of breast cancer and those with the variant that is associated with increased free radicals are at higher risk, but this study was the first to look at the effect of drinking tea. The study results showed that drinking tea significantly reduced the risk of breast cancer in women who had the *ACE* variant associated with higher levels of free-radical production and had no effect on those who had the low-free-radical-producing variant.

Preliminary studies have also found an association between the same high-free-radical-producing variant and an increased risk of other cancers, such as stomach cancer and prostate cancer, although the risk-lowering effects of drinking tea haven't yet been tested with relation to these cancers.

These findings demonstrate that when it comes to breast cancer, your genes will determine whether tea drinking has protective properties *for you*.

There is also good news for coffee lovers. Coffee appears to substantially lower the risk of breast cancer in those who have a specific variant in their *CYP1A2* gene, which produces an enzyme that is responsible for metabolizing almost all the caffeine we drink. When people drink coffee on a regular basis, the enzyme actually senses this and ramps up its activity so that it is able to metabolize more caffeine in a shorter period of time. However, a variant in the *CYP1A2* gene not only prevents it from ramping up its activity when exposed to daily caffeine but also decreases the activity of its enzyme so that it isn't able to process caffeine as quickly. As a result, the body is consistently exposed to much higher levels of caffeine over longer periods of time.

Analyzing your genetic makeup can provide information as to whether drinking coffee or tea may decrease your risk of breast cancer.

Caffeine appears to protect against various types of cancer, including breast cancer, skin cancer, and liver cancer, by altering hormone levels in the body, acting as an antioxidant, and directly inhibiting the growth of cancer. Therefore, if one has the variant that allows the body to be exposed to more caffeine, drinking coffee can have a significant protective effect. If a woman has specific *BRCA1/2* variants plus the *CYP1A2* variant, *and also* drinks 4 or more cups of coffee per day on a regular basis, her risk of breast cancer is reduced by about 60 percent. But for those who don't have the *CYP1A2* variant, coffee drinking has no protective effect.

■ PREGNANCY AND BREASTFEEDING

The risk of breast cancer is decreased in women who give birth to their first child within 16 years of their first menstrual cycle, and the risk of breast cancer can be further reduced in women of any age who breastfeed their baby. Both of these factors protect against breast cancer by causing changes in the structure of the glands and cells in the breast that make them less susceptible to cancer.

Giving birth for the first time earlier in life and breastfeeding for longer than a year have both been shown to decrease the risk of breast cancer in women with *BRCA1* variants by almost 50 percent. Unfortunately, however, no such benefit has been observed in women with *BRCA2* variants, most likely because the variants in *BRCA1* and *BRCA2* predispose women to different *subtypes* of breast cancer.

In the United States, 74 percent of women start off breastfeeding, but only 43 percent are still breastfeeding at 6 months and only 23 percent at 12 months. If, however, a new mother with the *BRCA1* variant understands the benefit to be derived from breastfeeding for a longer period of time, she might decide to breastfeed for longer than she would have otherwise, such as for a year or longer.

Breastfeeding has been shown to significantly decrease a woman's risk of breast cancer.

While initial studies found that women who have their first child by the age of 35 are at reduced risk of breast cancer, more recent studies have taken this a step further and concluded that the important value is not absolute age but rather the time that elapses between a woman's getting her first period and having her first child. The data suggest that the risk for breast cancer increases by about 50 percent when the time frame between first period and first birth is 16 years or more. There are two reasons for this: First, the glands in their breasts haven't undergone the protective changes that occur with pregnancy and breastfeeding, and, second, the breast tissue has been constantly exposed, without a break, to monthly increases in hormone levels that occur throughout the menstrual cycle. Recent evidence also suggests that having a child earlier in life may decrease the number of stem cells in the breast, and because stem cells may be among the first to turn cancerous, reducing their overall number may be one of the reasons having a child earlier in life reduces the risk of breast cancer.

While childbirth decreases a woman's overall lifetime risk of breast cancer, especially later in life, her risk is actually *increased* during pregnancy and for the 10 years following the birth of her first child. This initial spike in breast cancer risk most likely occurs because pregnancy exposes a woman's body to a flood of hormones, including estrogen and progesterone, that are directly related to cancer growth and also because it induces immunologic changes to the breast tissue that appear to make it more susceptible to cancer over the short term. Furthermore, some women may have cancer cells already developing at the time they become pregnant, and the increased hormone levels with pregnancy provide an environment that stimulates their growth.

Increased surveillance for breast cancer should be implemented for women predisposed to breast cancer during pregnancy and after they give birth.

This information is especially pertinent for women genetically predisposed to breast cancer, because even though pregnancy substantially decreases her overall risk in the long term, it may increase her risk in the short term. Therefore, I usually recommend that women who are genetically predisposed to breast cancer have a screening breast examination along with an ultrasound of the breast when they find out they are pregnant and then again halfway through their pregnancy. They should also have an MRI of the breast immediately after pregnancy, then every 6 months for 10 years, and annually thereafter. While it might seem prudent to wait until after a woman gives birth to conduct screening, if she does have breast cancer, postponing the diagnosis and treatment by even 5 to 7 months can increase her risk of advanced cancer by as much as 150 percent.

■ BREAST EXAMINATIONS

There are two types of physical breast exams—those performed by the woman herself (referred to as self-exams) and those performed by a physician (referred to as physician exams). Many studies have investigated the true value of physical breast exams, with most large studies concluding that self-exams are not an effective method of helping save a woman's life. That said, however, preliminary studies focusing specifically on self-exams by women who are genetically predisposed to breast cancer have found that they *do* appear to be beneficial for detecting cancer at its earliest possible stage. The Cancer Genetics Studies Consortium, which was organized by the National Institutes of Health, recommends that this particular group of women begin monthly self-exams between the ages of 18 and 21 and that they should be instructed by their physician in the proper method of conducting the exam.

Physician exams have been found to be beneficial for increasing the detection rate of breast cancer even in women who are not genetically predisposed; up to 10 percent of breast cancers are detected by physician exams. For women who are genetically predisposed to breast cancer, the Cancer Genetics Studies Consortium recommends that physician exams be conducted every 6 to 12 months starting at age 25.

▪ MAMMOGRAMS, MRIs, AND ULTRASOUNDS

The goal of imaging-based screening exams is to detect cancer as early as possible. Although physical exams can detect cancers that are palpable, other techniques that image the inner layers of the body are able to detect very small cancers anywhere in the breast. The most commonly used imaging technique for detecting breast cancer is the mammogram, which is basically a specialized x-ray of the breast. Newer tests, such as ultrasound, which use inaudible sound waves, and MRIs, which use magnetism, are different from mammograms because they do not expose the patient to any radiation.

Limiting radiation exposure is extremely important for people who are genetically predisposed to cancer because it has the potential to *further* increase their risk of cancer. Many of the variants that increase a person's risk of cancer cause genes that are supposed to repair DNA damage (such as the *BRCA1/2*, *CHEK2*, and *ATM* genes) to malfunction. Because of this, the body isn't able to as effectively protect itself from DNA damage and since radiation causes DNA damage, anything that exposes the chest to radiation, such as CAT scans and even low-level radiation exposure from x-rays and mammograms, can further increase the risk of breast cancer. Because of this, I am extremely cautious about exposing children and adults who are predisposed to breast cancer to any form of radiation.

Radiation exposure from an x-ray can increase the risk of breast cancer for decades afterward. In people under the age of 20 who are genetically predisposed to breast cancer, radiation increases their risk of the disease by more than 200 percent; in people between the ages of 20 and 40 it increases their risk by more than 75 percent; and in people over 40 it increases their risk by 45 percent. One study even found that any benefit from mammograms conducted between the ages of 25 and 35 in women predisposed to breast cancer is completely negated by the increased risk caused by radiation.

If a woman is predisposed to breast cancer, avoiding radiation exposure throughout life can be a worthwhile preventive measure.

Because radiation is most harmful to children and teenagers, parents whose children have been tested and found genetically predisposed to breast or ovarian cancer should inform their pediatrician of the risk and to try to avoid x-rays and CAT scans except when absolutely necessary. For pregnant women and people under 20 who are genetically predisposed to breast or ovarian cancer, being proactive about limiting radiation exposure is an incredibly important preventive measure, but it cannot be instituted unless these at-risk people have been previously identified by genetic screening.

The best nonradiation method we currently have for breast cancer screening is the MRI, which actually has the ability to pick up some very early-stage breast cancers that even mammograms and ultrasounds may miss. However, because MRIs are so sensitive they may also detect a larger number of changes to the breast that require further testing (such as a biopsy) but turn out not to be cancer. Because of this, MRIs are usually used to screen for breast cancer only when a woman is genetically predisposed to the disease, although I foresee a time when breast cancer screening for all women will be MRI based. After reviewing the literature on breast cancer risk and radiation exposure, the American Cancer Society released updated guidelines in 2007 stating that any woman who has a 20 percent or higher lifetime risk of breast cancer should avoid radiation exposure and have MRI breast screening instead of mammograms. This applies to women with a genetic variant that substantially increases their risk of breast cancer and those with a family history of breast cancer. Unfortunately, it can be a fight at times to get some health insurance companies and HMOs to cover MRIs for breast cancer screening even when they are clearly indicated, such as if a woman is known to be predisposed to breast cancer due to *BRCA1/2* variants.

Currently, there is no routine screening for ovarian cancer because it is relatively rare. However, it is certainly *not* rare in women who are genetically predisposed to it. Therefore, when genetic screening detects that someone is at increased risk of ovarian cancer, I usually recommend an annual ultrasound of the ovaries performed by a gynecologist and a yearly blood test for CA-125. CA-125 is a protein detectable in the blood that becomes increasingly elevated with ovarian cancer, but because a few other conditions can also cause this protein to become elevated, it is not useful as an annual screening test on its own.

■ Genetically Tailored Chemoprevention and Chemotherapy

The preventive strategies discussed here focused on risk reduction through lifestyle modifications and screening exams. There are two other, more drastic strategies that can also be implemented. The first of these is using medications to block estrogen in the body. Whereas chemo*therapy* refers to the use of medication to treat an existing cancer, this strategy is referred to as chemo*prevention* because a medication is used to lower the risk of cancer that hasn't yet manifested.

Tamoxifen, the best known and most widely used of these medications, blocks the estrogen receptors throughout the body, thereby triggering a medication-induced menopause. Having much higher estrogen levels than men is one of the key reasons breast cancer is so much more prevalent in women, and blocking estrogen's function neutralizes its harmful effects. Because it blocks estrogen receptors, tamoxifen has side effects similar to those experienced during menopause, and it shouldn't be taken by women who are pregnant or become pregnant because it may harm the fetus.

In women with *BRCA1/2* variants, one option is to start taking tamoxifen at about the age of 35 to prevent cancer from occurring. Taken for 5 consecutive years, it has been shown to reduce the risk of breast cancer by about 40 percent. There are also some preliminary studies to suggest that tamoxifen can be used for prevention in women who have breast cancer–associated variants in other genes, such as *CHEK2*.

Tamoxifen, however, is not effective in all people because of genetic differences. Tamoxifen is considered a pro-drug because it is taken in an inactive form and has to be *converted* into its active form by the liver enzyme *CYP2D6*. If a variant in the *CYP2D6* gene affects the function of the enzyme, tamoxifen is either not converted to its active state or it's converted at a much slower rate, making it less effective or outright ineffective for preventing and treating breast cancer.

Genetic screening can assess the effectiveness of tamoxifen prior to it being prescribed.

Variants within the *CYP2D6* gene that are involved in tamoxifen's effectiveness are found in 1 out of every 10 people throughout the world, meaning that they are actually quite common.

Tamoxifen is also sometimes used as a *treatment* for breast cancer, along with surgery. Because of this, genetic screening to determine whether a woman has any *CYP2D6* variants that affect tamoxifen can be used to tailor chemotherapy to the patient's genetic makeup. A case in point is the story of a 35-year-old woman who was diagnosed with a small breast cancer. After her surgery, and after discussion with her physician, she chose to be treated not with the full spectrum of chemotherapy but only with tamoxifen. About a year later, she visited a genetic counselor who conducted genetic screening that included the *CYP2D6* gene, which had not previously been checked. The counselor found that, based on the woman's genetic code, tamoxifen was likely to be *ineffective*. Unfortunately, not even a month after getting these results, the patient found out that her breast cancer had recurred and had already spread to her bones, meaning that the disease was now in an advanced stage. If genetic screening had been conducted before treatment with tamoxifen it is very likely that she would have been given a different medication, which would have been more effective in protecting her life.

It is also important to point out that other medications, including some selective serotonin reuptake inhibitors (SSRIs such as Prozac, Paxil, and Zoloft), prescribed to treat the hot flashes associated with tamoxifen, can also interfere with the *CYP2D6* enzyme. Studies have shown that tamoxifen taken in conjunction with these SSRIs more than doubles the risk of cancer because of tamoxifen's decreased effectiveness. However, other SSRIs (such as Celexa, Lexapro, and Luvox) do not interfere with *CYP2D6* as much and do not alter tamoxifen's effectiveness.

When a person is found at increased risk of breast cancer, reflex analysis will automatically analyze variants within the *CYP2D6* gene to ascertain whether tamoxifen is likely to have normal or decreased effectiveness. The genetic report will include this and other pertinent information, including the need to avoid specific SSRIs if tamoxifen is ever prescribed; this makes the analysis truly actionable.

▪ SURGICAL PREVENTION

For many diseases, a small increase in risk can be thwarted by an equally modest preventive measure, such as a simple lifestyle modification. But as the lifetime risk of a disease increases, more drastic prevention may be deemed necessary. Nowhere is this more evident than with breast cancer prevention. Some *BRCA1/2* variants increase a woman's risk of breast or ovarian cancer so much that she is almost certain to get the disease at some point in her lifetime. For these people, one of the most aggressive preventive strategies in our arsenal is prophylactic surgery. Performing surgery on someone who is currently healthy to decrease his or her risk of potentially contracting a disease in the future is generally considered quite radical. But when the disease is as deadly as breast or ovarian cancer, this option is *not* as unthinkable as it might seem.

It is difficult for most people to understand what it is like to have lost multiple family members to breast cancer and know that you are at risk of the same fate. That knowledge sometimes propels women to minimize their risk as much as possible, and no way is more effective than to remove the organs that put them at risk: their breasts and their ovaries.

Surgery that removes both breasts reduces a person's risk of breast cancer by between 85 and 100 percent. The reason this preventive measure does not always completely remove the risk is that a small amount of breast tissue is always left behind in the chest wall and there is a rare chance that a cancer can form there. In a large study of women who underwent this type of preventive surgery, almost 4 percent were found to have the earliest stages of cancer already infiltrating their breast tissue, meaning that they already had undetected cancer.

The surgery has a number of potential complications, including its psychological impact. However, studies evaluating its psychological effects and quality-of-life ramifications have found that, looking back, women are usually satisfied with the surgery and the decision they made.

Because removal of the breasts does not decrease ovarian cancer risk, some women choose to remove both their breasts and their ovaries. An alternative is to remove both ovaries but not the breasts. Doing so reduces ovarian cancer risk by 85 to 100 percent and *also* lowers breast cancer risk by 55 to 70 percent. Overall, risk of death from both breast and ovarian cancer is lowered by 90 percent because removing the ovaries outright removes the major source of estrogen, along with a number of other potentially harmful hormones. However, because there is still some estrogen present in the body, tamoxifen may be used as an adjunct to reduce the risk of breast cancer even further.

Removal of the ovaries is most effective when performed before the age of 40 and is relatively ineffective if performed after 50, when women are generally producing little or no estrogen. However, removing the ovaries of a woman in her 20s, 30s, or 40s, thereby rendering her infertile, can have serious quality-of-life implications. Therefore, this type of surgery is performed only when a woman has determined that she won't have any more biological children or that she doesn't want to have children at all.

For women who have a *BRCA1/2* variant and have decided they won't have any additional children, I believe that removing the ovaries is an option that must be given serious thought, especially when compared to removing both breasts. Removal of both ovaries is a more straightforward surgical procedure that has significantly fewer complications and does not alter a woman's physical appearance. Patients who choose this option should continue to have regular breast screening exams, but many feel a great deal of comfort in knowing that their risk of breast cancer has been significantly reduced.

▪ Men and Breast Cancer

We can be empowered to conquer diseases such as breast cancer only if we understand our own personal risk early enough to implement a comprehensive preventive strategy. For this reason, I believe in screening newborns and children for risk of breast and ovarian cancer, and I also believe in screening boys and men. About 1 percent of all breast cancers occur in men, with the number of new cases increasing each year. And while men don't have ovaries, genetic variants within the *BRCA1/2* genes can also significantly increase their risk of prostate and testicular cancers.

GENETICALLY TAILORED PREVENTION FOR CHILDREN AND ADULTS PREDISPOSED TO BREAST AND/OR OVARIAN CANCER	
NO PREDISPOSITION TO BREAST OR OVARIAN CANCER	GENETIC PREDISPOSITION TO OR FAMILY HISTORY OF BREAST AND/OR OVARIAN CANCER
Physician-performed breast exam every 2–3 years in the 20s and 30s and then annually after 40	Avoid radiation (from x-rays, CAT scans, and mammograms) to the chest, abdomen, and pelvis throughout childhood and adulthood except when absolutely necessary

Mammogram every year starting at age 40	Self-examination of the breasts every month starting at age 18 after receiving instruction from a physician
No specific screening indicated for ovarian cancer	Physician-performed breast exam every year starting at age 25 or at the time of the first pregnancy, whichever comes first
	MRI screening of breasts every 12 months starting at age 25 or immediately after the birth of the first child, whichever comes first
	Ultrasound of the breast along with physician-performed breast exam conducted initially and then halfway through pregnancy; applicable for all pregnancies
	Women with *BRCA1* variants: consider having first child no later than 16 years after first period
	Breastfeed for at least 12 months
	Ultrasound of the ovaries every year starting at age 25
	Yearly CA-125 blood tests starting at age 25, used in conjunction with annual ultrasound
	Regular exercise, including intensive cardio, 4 or more times per week starting at age 15
	Maintaining a healthy body weight (BMI < 24) after menopause
	Increased daily consumption of green tea for people with specific genetic variants
	Increased daily consumption of coffee for people with specific genetic variants
	Avoidance of alcohol throughout life for those with specific genetic variants
	Consume at least 300 micrograms of folate daily for life
	Considering chemoprevention with tamoxifen for 5 years starting at age 35, depending on genetic analysis
	Consider surgery to remove ovaries and fallopian tubes after childbearing is completed
	Considering surgery to remove breasts, ovaries, and fallopian tubes after childbearing is completed

PREVENT AND PREVAIL

Cancer has taken too many lives, caused too much suffering, been the source of too much pain. We need new weapons in our war against this powerful enemy. We are learning, on a genetic level, why cancer forms, how it grows, and what enables it to spread, and we must now turn our knowledge against the disease. We must outsmart the genes that enable the formation of this ugliness within us. Through prediction and prevention we *can* defeat cancer.

Example of a Cancer-Related Panel

Breast and Ovarian Cancer Panel

- Breast cancer
- Ovarian cancer
- Effectiveness of tamoxifen
- Adverse reactions with tamoxifen
- Chemotherapy-induced leukemia
- Risk of blood clots
- Risk of blood clots with thalidomide
- Risk of bleeding
- Depression due to stressful life events
- Poor wound healing
- Effect of tea on breast cancer risk
- Effect of caffeine on breast cancer risk
- Effect of alcohol on breast cancer risk

For additional panels related to cancer, please visit www.Outsmart YourGenes.com/Panels.

Afterword
Our Genetic Manifest Destiny

MISCONCEPTION: Understanding my DNA and being able to integrate genetic information into healthcare is a long way off.

FACT: The genetic revolution has already begun, and it is having far-reaching effects on healthcare right now. We have already taken the first leaps forward and, yes, we can benefit from it *today.*

Throughout this book we've discussed how predictive medicine, based on genetic screening and prevention, can have a direct, useful application to everyone's health and well-being. The capabilities of genetic technology do not stop there, however, and in closing I would like to discuss some of the other uses we can apply it to now and in the near future.

NEXT-GENERATION PHARMACEUTICAL CLINICAL TRIALS AND GENETICALLY TAILORED MEDICATIONS

The rapidly emerging field of pharmacogenomics is revolutionizing not only how medications are prescribed but also how they are developed and tested.

As we've already discussed, pharmacogenomics integrates genetic testing and analysis into pharmacology by using genetic information to determine how and why medications act differently on different people. This insight can then be used to predict which medications will be most (or least) effective in treating a *particular* person for a *particular* ailment, to determine whether he or she is at increased risk of experiencing side effects, and to prescribe the optimal starting dose. Instead of prescribing the same medication at the same dose to everyone with the same ailment, pharmacogenomics allows us to treat each person as an individual.

Knowing how different individuals react to a particular drug is also of significant concern during clinical trials that test the safety and effectiveness of newly developed medications. Very often these trials uncover unexpected, harmful reactions, and if enough people have this adverse reaction, the medication will be deemed unsafe for the general population and the FDA will not approve it. In other cases, a medication may be found to be only borderline effective, and the company may, therefore, decide not to produce it or, again, it may not receive FDA approval.

By integrating genetic screening into the development and testing of medications, pharmaceutical companies can now more accurately identify those people for whom a medication will be most appropriate. Instead of discovering that a random subset of their clinical trial population is having a severe adverse reaction to a medication, the company can analyze the clinical trial population's genetic makeup to ascertain whether people who are having an adverse reaction also share the same genetic makeup. For example, if people with a specific group of genetic variants do not have the adverse reaction but people with another set of variants do, the pharmaceutical company can show the FDA that they have found a way to *differentiate* the people who will have adverse reactions from those who won't. The FDA is then much more likely to approve the medication because the prescribing information will indicate who should *not* take the drug.

In the past few years, many pharmaceutical companies have started to use pharmacogenomics in their clinical trials. While this primarily entails analyzing only a small subset of genes that are involved in medication metabolism, comprehensive genetic screening now allows companies to screen study populations for practically all known diseases. This is important because sometimes being a carrier of a variant that causes a recessive disease, such as sickle-cell anemia and cystic fibrosis, may cause slight changes to the molecular functioning of cells that could ultimately affect the efficacy of the drug or produce side effects. Comprehensive genetic screening adds a new dimension to pharmaceutical clinical trials because it enables the analysis of all genes.

Although it hasn't been done yet, comprehensive genetic testing and analysis is also capable of *resurrecting* trials of medications that have previously failed to gain FDA approval because of a severe adverse reaction or because the medication wasn't deemed effective. By creating very comprehensive genetic profiles of all study participants and running an advanced analysis of the results, we can attempt to identify patterns in their genetic makeup that are responsible for the adverse reaction or the ineffectiveness of the drug. Using this information, it will be possible to genetically tailor a new clinical trial that includes only people who are likely to benefit. FDA approval would then be contingent on prescribing the drug only to individuals for whom it is not genetically contraindicated, thereby creating a genetically tailored medication.

The immense benefit of using genetic screening to test the safety and effectiveness of medications has already led the FDA, in a process known as retrofitting, to change their prescribing indications for certain medications already on the market. Retrofitting entails adding new recommendations for genetic screening of a patient *before* the medication is prescribed to avoid the drug being taken by people for whom it is genetically contraindicated.

An example of retrofitting is the medication abacavir, which was approved by the FDA in 1998 to slow the progression of HIV infection in HIV-positive individuals. It soon became evident, however, that as many as 8 percent of people were having a severe adverse, sometimes even fatal, reaction to the medication. A number of studies then found that the reaction was highly associated with a specific genetic variant, and a subsequent study found that conducting genetic screening before prescribing abacavir could reduce the incidence of these reactions by 100 percent, meaning that it completely eliminated the risk. Because of these findings, in 2008, the FDA retrofitted the labeling of abacavir so that its prescribing instructions now recommend that physicians perform genetic screening before prescribing it.

MEDICATIONS ASSOCIATED WITH KNOWN GENETIC VARIANTS THAT DICTATE EFFECTIVENESS, ADVERSE REACTIONS, AND/OR DOSING	
CLASS OF MEDICATION(S)	**USED IN THE TREATMENT OR PREVENTION OF**
Antibiotics	Bacterial infection
Antivirals	HIV/AIDS and hepatitis C
Statins	High cholesterol
Beta-blockers, angiotensin-converting enzyme inhibitors, angiotensin receptor blockers, diuretics	High blood pressure and heart failure
Blood thinners (warfarin, aspirin, Plavix)	Risk of blood clots, heart attack, stroke
Stomach acid blockers (proton pump inhibitors)	Stomach ulcers and acid reflux
Chemotherapy and chemoprevention (tamoxifen, 5-fluorouracil, paclitaxel, doxorubicin, vinblastine, vincristine, irinotecan, cisplatin, oxaliplatin, rituximab, prednisone, Herceptin)	Cancer (colorectal, breast, ovarian, leukemia, lymphoma, pancreatic, stomach, melanoma, lung, testicular, head and neck, sarcoma, bladder, and endrometrial)
Antidepressants (selective serotonin reuptake inhibitors, also known as SSRIs)	Depression, anxiety, eating disorders
Stimulants (Ritalin and Adderall)	Attention-deficit/hyperactivity disorder
Antiarrhythmias	Heart arrhythmias
Antipsychotics	Schizophrenia, bipolar disorder, psychosis
Antiepileptics	Epilepsy and seizure disorders
Antituberculosis	Prevention and treatment of tuberculosis

Inhaled lung medications	Asthma and chronic obstructive pulmonary disease
Immunosuppressants (azathioprine, 6-mercaptopurine, D-penicillamine, cyclosporine)	Organ transplants and autoimmune diseases (Crohn's disease, ulcerative colitis, multiple sclerosis, and rheumatoid arthritis)
Growth hormone	Abnormally small stature
General anesthesia and neuromuscular blockers	Used during surgical procedures

GENE THERAPY AND GENETIC ENGINEERING

Gene therapy and genetic engineering are ways to *modify* and *correct* our genetic makeup.

Up to this point we've been discussing the fact that once we become aware of our genetic risk for any particular disease, we can modify our nongenetic risk factors to lower our overall risk. This strategy is based on the assumption that we cannot modify genetic risk.

Gene therapy and genetic engineering, however, provide us with the ability to do just that. Gene therapy involves isolating normal genes in a laboratory and inserting them into an abnormal cell so that the abnormality is corrected. Genetic engineering involves changing the genetic makeup itself.

Quite often a variant causes harm by preventing its gene from functioning at full capacity, or at all. When that happens, inserting a normally functioning gene through gene therapy cures the problem. As an example of gene therapy, a genetic variant causes the *HBB* gene to produce an abnormal protein, leading to sickle-cell anemia. To test the effectiveness of gene therapy for this disease, scientists studied mice that had sickle-cell anemia. Researchers inserted copies of normal *HBB* genes into the cells of the diseased mice and the mice were cured. The normal gene that was inserted was able to act in place of the malfunctioning gene.

While this may sound like a relatively straightforward procedure, it is an extremely intricate process. To insert a new gene into the cells of a living person or animal, scientists have routinely used a natural delivery mechanism for genetic material: a virus. A virus usually acts by binding to a cell and injecting its genetic material into the cell. Therefore, a virus can be thought of as nature's own molecular-size injector. Scientists have been able to strip a virus of all the genes it normally uses to replicate and cause harm while leaving intact all the genes necessary for it to attach and inject its genetic material into a cell. This basically creates a zombie virus that can do nothing but inject genes into cells. The scientists then insert the gene they are using for therapy into the virus. Some of the cells from a person or animal are removed (such as by taking a blood specimen) and the cells are then exposed to the virus. Alternatively, the person may be exposed to the altered virus by inhalation or injection, after which the virus does its thing, attaches to the cells it normally infects, and injects its altered genetic material into the cell.

The virus then dies but the gene lives on inside the cell and starts functioning as a normal gene would. So even though the cell may contain its original malfunctioning gene, it also contains a normal gene, which is functioning correctly, so the cell starts functioning like a normal cell, and the person or animal is cured.

Unfortunately, ensuring that all of these steps work correctly is much more difficult than theorizing about it. Think about the genes on a chromosome as houses on a street. Your house is gene A, and your neighbor's house is gene B. When a new gene is inserted into a cell, it is as if someone new were moving into your neighborhood. If there is enough room, a new house can be built between your house and your neighbor's house. But if there isn't enough space between the houses, the new one will end up encroaching on land that is owned by you or your neighbor. The same thing can happen with the new gene; when it is integrated into a person's genetic code, it may be inserted into a part of the code that already contains another gene and may, therefore, interrupt the proper functioning of the current tenant. If this occurs, the gene therapy could create more harm than good.

Recently, great strides have been made in understanding why this occurs and finding ways to make sure it doesn't happen. For example, one gene therapy trial currently in progress is attempting to cure a disease that causes childhood blindness. Initial results are extremely promising, showing some return of sight with limited side effects. Gene therapy research is moving forward at a fast pace, and many of the diseases we've discussed throughout this book have the potential to one day be treated this way.

Genetic engineering is another extremely powerful technology of the near future. Gene therapy attempts to correct a problem by inserting a normal copy of the gene into the cell; genetic engineering would be able to actually change the genetic makeup of the gene itself. So, for example, if a disease occurred because one of the letters of gene X was a G when it was supposed to be an A, gene therapy would insert a copy of gene X that had the letter A while genetic engineering would actually modify the existing gene by changing the G to an A.

This technology is far more complicated than either genetic screening or gene therapy, but we're quickly approaching the day when it will be a reality.

Currently, a significant amount of research is being conducted on the genetic engineering of animals and plants. Instead of changing specific letters in a gene, researchers have been testing ways of inserting completely new genes from one species to another in a process known as *transgenics*. For example, genes from fish and fireflies that are responsible for luminescence have been successfully inserted into a wide range of plants and animals. The result is known as a genetically modified organism (GMO). You may be amazed to discover that luminescent cats, rabbits, pigs, monkeys, and trees have *already* been created, and they're alive right now in laboratories (just do an Internet search for "glowing cat," "glowing pig," and "glowing monkey" to see some amazing pictures). Yes, cute little bunnies that glow in the dark and monkeys that glow green *do* exist, although, much to my disappointment, we haven't yet been able to make pigs fly. That, however, is also a possibility.

Once research gets to the point where these technologies become applicable to a wide range of diseases, medicine will once more be transformed, and it will be time for me to write a new book titled *Change Your Genes*. Until that time, however, we are still empowered by current genetic technology to outsmart our genes in ways that were never before possible.

ANIMAL GENETIC TESTING

The structure and functioning of DNA and the four letters it contains are the same for all species. It is the *combination* of these letters that is distinctive, not the letters themselves. Because of this, the genetic testing technology we use for humans, such as arrays and full genome sequencing, can also be used for other species.

However, the key, once again, is not the actual laboratory testing but the genetic analysis that occurs after the testing. And the technologies that make advanced genetic analysis possible in humans are also fully applicable throughout the animal and plant kingdoms. Technologies such as the disease matrix, reflex analysis, and panels are all fully applicable, and I have already started to adapt this technology to many different species.

In Chapter 6, we discussed the Pythia Approach, which involves analysis that takes into account the genetic makeup of two potential parents and, using this information, is able to predict the likelihood of diseases and traits in their future offspring. Using this same approach, we will soon be able to provide detailed information on which two animals when mated would be most likely to have offspring with desired traits and least likely to have disease.

Genetic screening can also be used to create a DNA fingerprint database that would allow for animals to be tracked. For example, if raw meat were found to be infected with dangerous bacteria, we could determine the DNA fingerprint of the animal by taking a sample of the meat and trace it back to determine exactly where the meat came from and what else may be contaminated. Instead of spending weeks or months tracking down the source, we can now use DNA fingerprinting of animals in much the same way crime-scene investigation uses it in humans—to track down the culprit in a matter of hours.

And just as genetic screening can be used in humans, we'll soon be able to use it to diagnose diseases and genetically tailor both prevention and treatments for our pets. For example, genetic screening of a dog may indicate that he has a higher than normal likelihood of having specific food allergies, thereby allowing a veterinarian to recommend food that the dog is more likely to tolerate.

We are invariably linked to the animals around us. Some animals, such as dogs and cats, provide companionship; some, such as cows, chickens, and pigs, provide nourishment; and others, such as horses, provide sporting entertainment. Comprehensive genetic screening will soon be capable of maximizing the rewards we gain from all of these relationships.

THERE'S A GENE FOR THAT

Throughout the book, we've discussed a wide variety of diseases, and there are still many others, including Crohn's disease, arthritis, lupus, depression, and alcoholism, for which predictive medicine can provide risk information and genetically tailored prevention and treatment.

We can now apply predictive medicine to virtually all medical specialties. Even palliative medicine, which focuses on pain management and end-of-life care, may soon benefit from the insight provided by genetics. For example, Dr. Bruce Miller Jr., a Harvard-trained palliative care physician who practices in California, is working on ways to apply predictive medicine to his field. He believes that genetic screening has the potential to allow him to genetically tailor medications for more effective pain management.

Over the next few years, as more and more areas of medicine benefit from predictive medicine and genetic screening, we will see a time when medicine is fully personalized and focused on prevention. Genetic technology allows us to be proactive rather than reactive in our fight against disease.

PREDICTIVE MEDICINE AND THE TECHNOLOGICAL SINGULARITY

THROUGHOUT HISTORY, our civilization has encountered numerous species-level events, some of which have pushed the human

continued

race to the brink of extinction while others have been necessary for our continued survival and have enabled our species to
flourish.

Just as archaeologists are able to provide clues about ancient
cultures by studying physical remains, genetic researchers are
able to study human DNA in order to glean information about
the nature of human existence tens of thousands of years ago.
By examining DNA from people all around the world, genetic
researchers working with *National Geographic* were able to
peer back into human history and examine our distant lineage.
As a result of these studies, it is now apparent that about
70,000 years ago the human population throughout the entire
world totaled fewer than 2,000 people. What this means is that
at one point we were an endangered species!

At that time, everyone lived in Africa because the rest of the
world hadn't been inhabited yet, and it appears that harsh
environmental conditions, such as widespread droughts and
global cooling due to a massive volcanic eruption, contributed
to the population decline. The human race was fighting for its
survival, and if our ancestors hadn't won that battle we would
not exist today.

Around this time something occurred that changed our species forever. Small groups of people started to band together
and form communities. These communities allowed for the division of labor. Individuals began to specialize in the tasks they
did best, such as hunting, gathering, cooking, shelter building,
protection, making the tools necessary for each of those tasks,
and so on. Each individual benefited so significantly from this
community-based life that humans were brought back from the
brink of extinction. Because of this, the formation of communities was a species-level event and a major turning point for our
civilization.

Other equally significant species-level events include:

- The ability to create fire, which allowed us to cook food, survive extremes of temperature, and make many different types of tools.
- The development of complex spoken language, writing, and much later the printing press, all of which allowed for the acquisition and transfer of knowledge from one generation to the next.
- The invention of the wheel and the sail, both of which were necessary for the development of trade and exploration. And much later, the invention of the engine.
- The first agricultural revolution (known as Neolithic Revolution), which literally allowed us to put down roots and move from a nomadic, hunter/gatherer way of life to a community-based farming existence.
- Empiricism and the rise of the evidence-based sciences, which allowed us to move away from mysticism and gain a deep understanding of nature.
- The germ theory of disease and, much later, the invention of the microscope, which became cornerstones of science and medicine and allowed us to understand there are worlds beyond that which we can see with the naked eye.
- Free trade, which allowed for tremendous value creation through the division of labor that, with its integration with democracy, became based on freedom, meritocracy, and an individual's inalienable rights.
- The Industrial Revolution, a time of tremendous technological and economic progress that physically connected all parts of the world through trade and travel, signaling the beginning of a global civilization.
- Sanitation and, later, hygiene, which reduced many common causes of death and significantly increased human life expectancy.

continued

- The Information Age, now under way, which has allowed for instantaneous global dissemination of, and access to, an unprecedented amount of information.
- The genetic revolution, now in its infancy, which will allow us to prevent disease before it even manifests and provides the potential to not only preserve our health and wellness but also greatly extend human longevity.

Species-level events appear to be accelerating at an exponential rate. While the time between each of these events may once have been tens of thousands of years, it is now only a few decades. And, in the near future, these events will most likely occur with even greater frequency because each successive species-level event utilizes and builds on those that have come before it, thereby reducing the time it takes for the next event to occur. This means that we are quickly approaching a time when significant advancements, such as great leaps in increasing human longevity, may be made virtually back-to-back.

This concept is based on the law of accelerating returns and is known as technological singularity, an idea made popular by Raymond Kurzweil. The technological singularity is that time at which our civilization has obtained so much technological knowledge that we are able to make significant advancements in relatively short periods of time. So instead of the next species-level event taking 50 years or even 10 years, it might take just a year. Just imagine all the advancements that are now possible because the Information Age has made all of the world's knowledge accessible to virtually anyone, anywhere, at any time.

Tremendous advances in genetics, including predictive medicine, gene therapy, and genetic engineering, are ushering in our newest species-level event: the genetic revolution.

THE GENETIC REVOLUTION

Healthcare costs for the entire world are spiraling out of control. In the United States alone, annual healthcare spending is now more than $2.5 trillion and increasing by approximately 7 percent *every year*. This represents nearly 20 percent of the U.S. gross domestic product (GDP), with healthcare spending in Canada and Europe also being very high. More than 75 percent of the cost of healthcare is attributable to chronic diseases, with heart disease in the United States accounting for around $450 billion; smoking, $193 billion; diabetes, $174 billion; arthritis, $128 billion; obesity, $117 billion; and cancer, $90 billion. The vast majority of this money goes to disease management once the disease has already manifested.

A revolution is needed because our current healthcare model is unsustainable. Just as we need a paradigm shift from an unsustainable reliance on fossil fuels to renewable forms of energy, so too does the next generation of healthcare require a shift from our dependence on reactionary medicine to a more sustainable proactive approach. Predictive medicine provides the answer because it is based not on propagating the traditional one-size-fits-all reactionary model but on personalized prevention. For our healthcare system to survive, we can no longer rely solely on our ability to reactively treat and must now focus on our power to *predict and prevent*. Not only does our own personal health depend on it but so does the very survival of the entire global healthcare system.

We stand on the precipice of a new species-level event that will radically change healthcare forever. The genetic revolution is now all about one single individual—*you*.

This is just the beginning.

Epilogue
An Update on the Predictive Medicine Revolution

I n the few short months since the hardcover edition of *Outsmart Your Genes* was published, significant advances have been made in the rapidly emerging field of predictive medicine.

For example, based on what we now know about the link between exposure to radiation during medical tests such as x-rays, even in childhood, and an increased risk of breast cancer in individuals genetically predisposed to the disease, physicians have begun to alter the way they practice medicine. Many physicians are now starting to order MRIs instead of x-rays, CAT scans, or mammograms for these patients. As more people elect to have genetic screening for themselves and their children, we will be able to identify those who are most at risk from radiation exposure during medical procedures and, therefore, limit exposure throughout the person's life.

We have also made progress in our battle against Alzheimer's disease. Because of physicians' growing awareness of the effects that head injury can have on the development of Alzheimer's and other forms of dementia, the U.S. government has become involved in what is now a national discussion of the problem. In the fall of 2009, the House Judiciary Committee convened hearings on the incidence of brain injury among participants in contact sports on all levels, including high school, college, and professional, specifically including the NFL.

Because genetic screening can identify those people who are at considerable risk of Alzheimer's and dementia after head injury, we can now alert them—and their parents, in the case of children—to the problem so that they can take the steps necessary to mitigate that risk. Again, this is an example of how identifying people with specific genetic variants can help prevent the onset of disease.

Finally, one of the most exciting advances is coming in the area of infectious diseases, specifically HIV/AIDS. We know that the CCR5 gene produces a protein that forms the scaffolding on the outside of immune cells that HIV clings to in order to infect the cell. We now also know that a variant in CCR5 can cause the gene to malfunction so that it isn't able to produce any of its protein and, therefore, the immune cells lack their normal outer scaffolding. With nothing to grab onto, the HIV virus is unable to infect the cell. Because of this, people with two copies of this variant, making both copies of his or her CCR5 gene nonfunctional, will remain HIV negative even after exposure to the HIV virus. In these individuals, the virus will just circulate in the blood for a while before it eventually dies. People with this variant in only one of their pair of CCR5 genes can still become infected, but if they do become HIV positive, the disease will progress much more slowly before it becomes AIDS.

FIGHTING WEST NILE VIRUS WITH PREDICTIVE MEDICINE

WEST NILE VIRUS (WNV) is spread from mosquitoes to humans when someone is bitten by an infected mosquito. About 20 percent of people who are bitten by a WNV-infected mosquito develop symptoms of the disease. For this 20 percent, however, the symptoms can range from mild fever, headache, weakness, and joint pain to quite severe, including infection of brain tissue, long-term paralysis, and even death.

The first case in North America was diagnosed in 1999 in New York City, but the virus has existed for thousands of years

continued

throughout Africa and the Middle East. In fact, some medical historians postulate that it is to blame for the death of Alexander the Great, who died in 323 BC in Mesopotamia (now known as Iraq), as many of his symptoms were consistent with the disease and an ancient text describes an event preceding his death in which Alexander witnessed birds acting erratically, with some falling dead right in front of him.

The description of birds acting erratically and dying may have been indicative of a WNV outbreak, because birds can become infected and die from the virus just as humans can. This is why you may hear that researchers routinely test dead birds found during the summer, especially in large numbers, for WNV and alert the public to an outbreak so that people can take preventive measures.

Though people who have the variant in both copies of their CCR5 gene are usually immune to HIV infection, these same individuals are actually at greatly increased risk of infection and death if exposed to WNV because the immune system uses the CCR5 scaffolding to defend against infection by WNV. Therefore, a properly functioning CCR5 gene is crucial to fending off WNV. Numerous studies have found that the same CCR5 variant that makes people resistant to HIV infection also makes them much more susceptible to West Nile virus. And these individuals aren't simply more susceptible; their symptoms are also much more severe and, because of this, they have a substantially increased risk of dying from the disease.

Predictive medicine is now able to warn people who have this variant in their CCR5 gene that if there is ever an outbreak of WNV in their area, they need to take extra precautions to prevent infection. For example, they can protect themselves from mosquitoes by making sure to drain any stagnant water (which is where mosquitoes breed), wearing long-sleeved shirts and pants, using screens in open windows, making sure the screens are in good repair, limiting time spent outdoors, and applying bug repellent with a high percentage of a substance called DEET to their clothing and hair when they go outside.

Knowing that a patient is genetically extra-sensitive to WNV infection may also influence the medical care provided to him or her by a physician. If thought to be infected, the doctor may, for example, admit the patient to the hospital so that he or she can be more closely monitored. While there are no standard treatments available for WNV, if a person does become infected and is showing signs of severe disease and rapid deterioration, a medication called interferon may be used to try to prevent it from progressing further. While this is not yet an FDA-approved treatment, a small preliminary study has found interferon to be associated with rapid improvement in deteriorating WNV patients.

One of the most fascinating developments concerning the CCR5 gene occurred when Dr. Gero Hütter, a German physician, published a groundbreaking case report in the *New England Journal of Medicine* in 2009 concerning a patient whom he had cured of HIV infection. The patient, who was 40 years old, had been diagnosed with HIV 10 years before and was being treated with standard HIV medications. When the patient developed leukemia, however, Dr. Hütter decided to attempt a new procedure—transplanting him with stem cells taken from the bone marrow of a donor who was specially chosen after genetic testing showed that he had the variant that made both copies of his CCR5 gene nonfunctional. Dr. Hütter theorized that transplanting his HIV positive patient with stem cells from a donor who was immune to HIV might potentially cure the recipient's HIV infection. In essence, he was transplanting an immunity to HIV infection from one person to another.

First, the patient's own immune system was destroyed and then the donor's stem cells were transplanted into his bloodstream. These new stem cells survived and flourished and started to produce immune cells that didn't contain the normal CCR5 scaffolding. As a result, the HIV virus that had lived so long in Dr. Hütter's patient was unable to infect these new immune cells and the virus eventually died. After the transplant, testing showed that the patient had become HIV negative, and he stopped taking all his medications. After more than two years off the medications, and with repeated testing, the patient remains HIV negative.

Before widespread use of this procedure is attempted, however, more research has to be conducted to replicate the results in a greater number of people. Also, this type of stem cell transplant has a high mortality rate, so advancements have to be made in order to increase its safety. All that aside, Dr. Hütter's success has opened the door to powerful new research initiatives that are now not only investigating this transplant option but also searching for new medications and gene therapies that might inactivate either the CCR5 gene or its protein so as to mimic the effects of having the CCR5 variant. While curing a single patient may seem statistically insignificant, it is actually a monumental achievement because it shows that HIV infection can, in fact, be cured.

This is only one of the many possible cures that may soon be achieved through the combined use of genetic technology and stem cell transplants. If, for example, we can find a way to cause a stem cell to turn into a healthy brain cell, and we learn how to properly implant that new brain cell into the correct place in the brain, we may be able to cure Parkinson's disease. Similar cures may one day exist for diabetes, multiple sclerosis, Huntington's, and many other diseases.

These are just some of the newest advancements in the many ways genetic technology is changing the face of medicine and will continue to drive advancements in predictive medicine for the future. Thanks to the genetic revolution, we have truly entered a new era of medicine!

References

Outsmart Your Genes is based on a plethora of scientific research and contains more than 1,500 references, including research studies, journal articles, books, interviews, personal research, and other sources. This section contains an abridged list of references in order to show the immense scientific support for all of the groundbreaking topics discussed. The *unabridged* list of references can be found at www.OutsmartYourGenes.com/References. However, reading these references is by no means necessary in order to obtain full value from this book.

1. The Power of Predictive Medicine

Kolor, K., et al. (2009). "Health Care Provider and Consumer Awareness, Perceptions, and Use of Direct-to-Consumer Personal Genomic Tests, United States, 2008." *Genet Med* 11 (8): 595.

U.S. Food and Drug Administration. "Critical Path Initiative—Warfarin Dosing," 10/13/09: www.fda.gov/ScienceResearch/SpecialTopics/Critical PathInitiative/FacesBehindCriticalPath/ucm077473.htm.

Bauer, L. O., et al. (2007). "Variation in GABRA2 Predicts Drinking Behavior in Project MATCH Subjects." *Alcohol Clin Exp Res* 31 (11): 1780–1787.

Gordeeva, E. (1996). *My Sergei: A Love Story.* Warner Books, 255.

Goldschmidt-Clermont, P., et al. (1996). "Clues to the Death of an Olympic Champion." *Lancet* 347 (9018): 1833.

Weiss, E. J., et al. (1996). "A Polymorphism of a Platelet Glycoprotein Receptor as an Inherited Risk Factor for Coronary Thrombosis." *N Engl J Med* 334 (17): 1090–1094.

Bojesen, S. E., et al. (2003). "Platelet Glycoprotein IIb/IIIa Pl^{A2}/Pl^{A2} Homozygosity Associated with Risk of Ischemic Cardiovascular Disease and Myocardial Infarction in Young Men: The Copenhagen City Heart Study." *J Am Coll Cardiol* 42 (4): 661–667.

Zotz, R. B., et al. (2005). "Association of Polymorphisms of Platelet Membrane Integrins $\alpha_{IIb}\beta_3$ (HPA-1b/Pl^{A2}) and $\alpha_2\beta_1$ ($\alpha_2$807TT) with Premature Myocardial Infarction." *J Thromb Haemost* 3 (7): 1522–1529.

2. From Peas to Predictive Medicine

Mendel, G. (1865). "Versuche über Pflanzen-Hybriden." *Verhandlungen des naturforschenden Vereines in Brünn* 4: 3–47.

Watson, J. D., and Crick, F. H. (1953). "Molecular Structure of Nucleic Acids: A Structure for Deoxyribose Nucleic Acid." *Nature* 171 (4356): 737–738.

Atzmon, G., et al. (2005). "Biological Evidence for Inheritance of Exceptional Longevity." *Mech Ageing Dev* 126 (2): 341–345.

Willcox, B. J., et al. (2008). "FOXO3A Genotype Is Strongly Associated with Human Longevity." *Proc Natl Acad Sci USA* 105 (37): 13987–13992.

McNally, E. M. (2004). "Powerful Genes: Myostatin Regulation of Human Muscle Mass." *N Engl J Med* 350 (26): 2642–2644.

Papassotiropoulos, A., et al. (2006). "Common Kibra Alleles Are Associated with Human Memory Performance." *Science* 314 (5798): 475–478.

Yang, N., et al. (2003). "ACTN3 Genotype Is Associated with Human Elite Athletic Performance." *Am J Hum Genet* 73 (3): 627–631.

Devlin, B., et al. (1997). "The Heritability of IQ." *Nature* 388 (6641): 468–471.

Dick, D., et al. (2007). "Association of CHRM2 with IQ: Converging Evidence for a Gene Influencing Intelligence." *Behavior Genetics* 37 (2): 265–272.

Samson, M., et al. (1996). "Resistance to HIV-1 Infection in Caucasian Individuals Bearing Mutant Alleles of the CCR-5 Chemokine Receptor Gene." *Nature* 382 (6593): 722–725.

Blanpain, C. (2002). "CCR5 and HIV Infection." *Receptors Channels* 8 (1): 19–31.

Osborn, D. (1916). "Inheritance of Baldness." *J Hered* 7: 347–355.

Levy-Nissenbaum, E., et al. (2005). "Confirmation of the Association Between Male Pattern Baldness and the Androgen Receptor Gene." *Eur J Dermatol* 15 (5): 339–340.

Hillmer, A. M., et al. (2005). "Genetic Variation in the Human Androgen Receptor Gene Is the Major Determinant of Common Early-Onset Androgenetic Alopecia." *Am J Hum Genet* 77 (1): 140–148.

Bachner-Melman, R., et al. (2005). "Link Between Vasopressin Receptor AVPR1A Promoter Region Microsatellites and Measures of Social Behavior in Humans." *J Individ Dif* 26 (1): 2–10.

Bachner-Melman, R., et al. (2005). "AVPR1a and SLC6A4 Gene Polymorphisms Are Associated with Creative Dance Performance." *PLoS Genet* 1 (3): e42.

Lee, H.-J., et al. (2007). "Allelic Variants Interaction of CLOCK Gene and G-Protein Beta-3 Subunit Gene with Diurnal Preference." *Chronobiology International* 24 (4): 589–597.

Partonen, T., et al. (2007). "Three Circadian Clock Genes Per2, Arntl, and Npas2 Contribute to Winter Depression." *Ann Med* 39 (3): 229–238.

Zivelin, A., et al. (2006). "Prothrombin 20210G-A Is an Ancestral Prothrombotic Mutation That Occurred in Whites Approximately 24,000 Years Ago." *Blood* 107 (12): 4666–4668.

Lindqvist, P. G., et al. (2001). "Improved Hemoglobin Status and Reduced Menstrual Blood Loss Among Female Carriers of Factor V Leiden: An Evolutionary Advantage?" *J Thromb Haemost* 86 (4): 1122–1123.

Elisabeth, R. P., et al. (2007). "Risk of Venous Thrombosis: Obesity and Its Joint Effect with Oral Contraceptive Use and Prothrombotic Mutations." *Br J Haematol* 139 (2): 289–296.

Levy, S., et al. (2007). "The Diploid Genome Sequence of an Individual Human." *PLoS Biology* 5 (10): e254.

Eid, J., et al. (2009). "Real-Time DNA Sequencing from Single Polymerase Molecules." *Science* 323 (5910): 133–138.

Drmanac, R., et al. (2009). "Human Genome Sequencing Using Unchained Base Reads on Self-Assembling DNA Nanoarrays." *Science* doi: 10.1126/science.1181498.

4. Become Informed About Genetic Screening

(2008). "Direct-to-Consumer Genetic Tests: Flawed and Unethical." *Lancet Oncol* 9 (12): 1113.

Heshka, J. T., et al. (2008). "A Systematic Review of Perceived Risks, Psychological and Behavioral Impacts of Genetic Testing." *Genetics in Medicine* 10 (1): 19–32.

Chao, S., et al. (2008). "Health Behavior Changes After Genetic Risk Assessment for Alzheimer Disease: The REVEAL Study." *Alzheimer Dis Assoc Disord* 22 (1): 94–97.

Cassidy, M. R., et al. (2008). "Comparing Test-Specific Distress of Susceptibility Versus Deterministic Genetic Testing for Alzheimer's Disease." *Alzheimers Dement* 4 (6): 406–413.

Aspinwall, L. G., et al. (2008). "CDKN2A/p16 Genetic Test Reporting Improves Early Detection Intentions and Practices in High-Risk Melanoma Families." *Cancer Epidemiol Biomarkers Prev* 17 (6): 1510–1519.

Meigs, J. B., et al. (2008). "Genotype Score in Addition to Common Risk Factors for Prediction of Type 2 Diabetes." *N Engl J Med* 359 (21): 2208–2219.

Lyssenko, V., et al. (2008). "Clinical Risk Factors, DNA Variants, and the Development of Type 2 Diabetes." *N Engl J Med* 359 (21): 2220–2232.

Gever, J. "Genetic Screening Offers Little Help for Diabetes Prediction." *MedPage Today*, 11/19/08: www.medpagetoday.com/Nephrology/Diabetes/11842.

Gardner, A. "Genetic Testing No Real Help in Predicting Type 2 Diabetes." *US News & World Report*, 11/19/08.

5. Do These Genes Make Me Look Fat?

Bray, M. S., et al. (2009). "The Human Gene Map for Performance and Health-Related Fitness Phenotypes: The 2006–2007 Update." *Med Sci Sports Exerc* 41 (1): 35–73.

Walley, A. J., et al. (2006). "Genetics of Obesity and the Prediction of Risk for Health." *Hum Mol Genet* 15 (Spec No. 2): R124–R130.

Arkadianos, I., et al. (2007). "Improved Weight Management Using Genetic Information to Personalize a Calorie Controlled Diet." *Nutr J* 6 (1): 29.

National Center for Chronic Disease Prevention and Health Promotion. (2009). "Obesity Halting the Epidemic by Making Health Easier." Centers for Disease Control and Prevention and the Department of Health and Human Services.

Thorpe, K. E., et al. (2009). "Weighty Matters: How Obesity Drives Poor Health and Health Spending." National Business Group on Health.

Chopra, M., et al. (2002). "A Global Response to a Global Problem: The Epidemic of Overnutrition." *Bull World Health Organization* 80: 952–958.

Harvey-Berino, J., et al. (2001). "Does Genetic Testing for Obesity Influence Confidence in the Ability to Lose Weight? A Pilot Investigation." *J Am Diet Assoc* 101 (11): 1351–1353.

Frayling, T. M., et al. (2007). "A Common Variant in the FTO Gene Is Associated with Body Mass Index and Predisposes to Childhood and Adult Obesity." *Science* 316: 889–894.

Thorleifsson, G., et al. (2009). "Genome-Wide Association Yields New Sequence Variants at Seven Loci That Associate with Measures of Obesity." *Nat Genet* 41 (1): 18–24.

Van Vliet-Ostaptchouk, J. V., et al. (2008). "Polymorphisms of the TUB Gene Are Associated with Body Composition and Eating Behavior in Middle-Aged Women." *PLoS ONE* 3 (1): e1405.

Vaisse, C., et al. (2000). "Melanocortin-4 Receptor Mutations Are a Frequent and Heterogeneous Cause of Morbid Obesity." *J Clin Invest* 106 (2): 253–262.

Chambers, J. C., et al. (2008). "Common Genetic Variation Near MC4R Is Associated with Waist Circumference and Insulin Resistance." *Nat Genet* 40 (6): 716–718.

Cauchi, S., et al. (2009). "Combined Effects of MC4R and FTO Common Genetic Variants on Obesity in European General Populations." *J Mol Med* 87 (5): 537–546.

Moreno-Aliaga, M. J., et al. (2005). "Does Weight Loss Prognosis Depend on Genetic Make-up?" *Obes Rev* 6 (2): 155–168.

Arkadianos, I., et al. (2007). "Improved Weight Management Using Genetic Information to Personalize a Calorie Controlled Diet." *Nutr J* 6: 29.

Seip, R., et al. (2008). "Physiogenomic Comparison of Human Fat Loss in Response to Diets Restrictive of Carbohydrate or Fat." *Nutr Metab* 5: 4.

Retey, J. V., et al. (2007). "A Genetic Variation in the Adenosine A2A Receptor Gene (ADORA2A) Contributes to Individual Sensitivity to Caffeine Effects on Sleep." *Clin Pharmacol Ther* 81 (5): 692–698.

Fogelholm, M., et al. (2007). "Sleep-Related Disturbances and Physical Inactivity Are Independently Associated with Obesity in Adults." *Int J Obes* 31 (11): 1713–1721.

Goyenechea, E., et al. (2006). "Weight Regain After Slimming Induced by an Energy-Restricted Diet Depends on Interleukin-6 and Peroxisome-Proliferator-Activated-Receptor-γ2 Gene Polymorphisms." *Br J Nutr* 96 (5): 965–972.

Fox, A. L. (1932). "The Relationship Between Chemical Constitution and Taste." *Proc Nat Acad Sci* 18 (1): 115–120.

Norman, B., et al. (2009). "Strength, Power, Fiber Types, and mRNA Expression in Trained Men and Women with Different ACTN3 R577X Genotypes." *J Appl Physiol* 106 (3): 959–965.

Massidda, M. (2009). "Association Between the ACTN3 R577X Polymorphism and Artistic Gymnastic Performance in Italy." *Genet Test Mol Biomarkers* 13 (3): 377–380.

Santiago, C., et al. (2008). "ACTN3 Genotype in Professional Soccer Players." *Br J Sports Med* 42 (1): 71–73.

Yang, N. (2003). "ACTN3 Genotype Is Associated with Human Elite Athletic Performance." *Am J Hum Genet* 73 (3): 627–631.

Gayagay, G., et al. (1998). "Elite Endurance Athletes and the ACE I Allele: The Role of Genes in Athletic Performance." *Hum Genet* 103 (1): 48–50.

Tsianos, G., et al. (2004). "The ACE Gene Insertion/Deletion Polymorphism and Elite Endurance Swimming." *Eur J Appl Physiol* 92 (3): 360–362.

Henderson, J., et al. (2005). "The EPAS1 Gene Influences the Aerobic–Anaerobic Contribution in Elite Endurance Athletes." *Hum Genet* 118 (3): 416–423.

Cherkas, L. F., et al. (2008). "The Association Between Physical Activity in Leisure Time and Leukocyte Telomere Length." *Arch Intern Med* 168 (2): 154–158.

Rubio, J. C., et al. (2005). "Frequency of the C34T Mutation of the AMPD1 Gene in World-Class Endurance Athletes: Does This Mutation Impair Performance?" *J Appl Physiol* 98 (6): 2108–2112.

Muniesa, C. A., et al. (2008). "World-Class Performance in Lightweight Rowing: Is It Genetically Influenced? A Comparison with Cyclists, Runners and Non-athletes." *Br J Sports Med* doi: bjsm.2008.051680.

Andreu, A. L., et al. (1999). "Exercise Intolerance Due to Mutations in the Cytochrome b Gene of Mitochondrial DNA." *N Engl J Med* 341 (14): 1037–1044.

Todd, T. (1992). "A History of the Use of Anabolic Steroids in Sport." *Sport and Exercise Science: Essays in the History of Sports Medicine*, ed. Jack W. Berryman and Roberta J. Park (Champaign, IL: University of Illinois Press), 330.

Schulze, J. J., et al. (2009). "Substantial Advantage of a Combined Bayesian and Genotyping Approach in Testosterone Doping Tests." *Steroids* 74 (3): 365–368.

Schulze, J. J., et al. (2008). "Doping Test Results Dependent on Genotype of Uridine Diphospho-Glucuronosyl Transferase 2B17, the Major Enzyme for Testosterone Glucuronidation." *J Clin Endocrinol Metab* 93 (7): 2500–2506.

Hsiao, D.-J., et al. (2009). "Weight Loss and Body Fat Reduction Under Sibutramine Therapy in Obesity with the C825T Polymorphism in the GNB3 Gene." *Pharmacogenet Genomics* 19 (9): 730–733.

Potoczna, N., et al. (2004). "Gene Variants and Binge Eating as Predictors of Comorbidity and Outcome of Treatment in Severe Obesity." *J Gastrointest Surg* 8 (8): 971–982.

Sesti, G., et al. (2005). "Impact of Common Polymorphisms in Candidate Genes for Insulin Resistance and Obesity on Weight Loss of Morbidly Obese Subjects After Laparoscopic Adjustable Gastric Banding and Hypocaloric Diet." *J Clin Endocrinol Metab* 90 (9): 5064–5069.

Chen, H. H., et al. (2007). "Ala55Val Polymorphism on UCP2 Gene Predicts Greater Weight Loss in Morbidly Obese Patients Undergoing Gastric Banding." *Obesity Surgery* 17 (7): 926–933.

6. Prospective Parents

Sagi, M., et al. (2009). "Preimplantation Genetic Diagnosis for BRCA1/2: A Novel Clinical Experience." *Prenat Diagn* 29 (5): 508–513.

Hathaway, F., et al. (2009). "Consumers' Desire Towards Current and Prospective Reproductive Genetic Testing." *J Genet Couns* 18 (2): 137–146.

Geifman-Holtzman, O., et al. (2008). "Prenatal Diagnosis: Update on Invasive Versus Noninvasive Fetal Diagnostic Testing from Maternal Blood." *Expert Rev Mol Diag* 8 (6): 727–751.

Caspi, A., et al. (2007). "Moderation of Breastfeeding Effects on the IQ by Genetic Variation in Fatty Acid Metabolism." *Proc Natl Acad Sci USA* 104 (47): 18860–18865.

Donovan v. Idant Laboratories. The United States District Court for the Eastern District of Pennsylvania, Civil Action #08–4075, 3/31/09 and 6/10/09.

(2009). "Sperm Bank Sued Under Product Liability Law." *New Scientist* 2703: 4.

MacArthur, D. "Sperm as a Defective Product: More Details." *Genetic Future* blog, 4/9/09: http://scienceblogs.com/geneticfuture/2009/04/sperm_as_a_defective_product_m.php.

Brenner, B., et al. (1999). "Thrombophilic Polymorphisms Are Common in Women with Fetal Loss Without Apparent Cause." *J Thromb Haemost* 82 (1): 6–9.

Coulam, C. B., et al. (2009). "Thrombophilic Gene Polymorphisms Are Risk Factors for Unexplained Infertility." *Fertil Steril* 91 (4 Suppl 1): 1516–1517.

Nelen, W. L., et al. (1997). "Genetic Risk Factor for Unexplained Recurrent Early Pregnancy Loss." *Lancet* 350 (9081): 861.

Poongothai, J., (2009). "Genetics of Human Male Infertility." *Singapore Med J* 50 (4): 336–347.

Shady, A. A., and Colby, B. R., et al. (2002). "Congenital Erythropoietic Porphyria: Identification and Expression of Eight Novel Mutations in the Uroporphyrinogen III Synthase Gene." *Br J Haematol* 117 (4): 980–987.

Sotos, J. G. (2008). *The Physical Lincoln Complete* (Scotts Valley, CA: CreateSpace).

7. Newborns and Children

Butz, A. M., et al. (1993). "Newborn Identification: Compliance with AAP Guidelines for Perinatal Care." *Clin Pediatr* 32 (2): 111–113.

De Pancorbo, M. M., et al. (1997). "Newborn Genetic Identification: A Protocol Using Microsatellite DNA as an Alternative to Footprinting." *Clinica Chimica Acta* 263 (1): 33–42.

Schwartz, P. J., et al. (1998). "Prolongation of the QT Interval and the Sudden Infant Death Syndrome." *N Engl J Med* 338 (24): 1709–1714.

Schwartz, P. J., et al. (2000). "A Molecular Link Between the Sudden Infant Death Syndrome and the Long-QT Syndrome." *N Engl J Med* 343 (4): 262–267.

Xue, Y., et al. (2009). "Human Y Chromosome Base-Substitution Mutation Rate Measured by Direct Sequencing in a Deep-Rooting Pedigree." *Curr Biol* 19 (17): 1453–1457.

Schwartz, P. J., et al. (1985). "The Idiopathic Long QT Syndrome: Pathogenetic Mechanisms and Therapy." *Eur Heart J* 6 (Suppl D): 103–114.

Freitag, C. M. (2007). "The Genetics of Autistic Disorders and Its Clinical Relevance: A Review of the Literature." *Mol Psychiatry* 12 (1): 2–22.

Rogers, S. (2000). "Interventions That Facilitate Socialization in Children with Autism." *J Autism Dev Disord* 30 (5): 399–409.

U.S. Food and Drug Administration. "Thimerosal in Vaccines," 11/16/09: www.fda.gov/BiologicsBloodVaccines/SafetyAvailability/VaccineSafety/UCM096228.

DeStefano, F. (2007). "Vaccines and Autism: Evidence Does Not Support a Causal Association." *Clin Pharmacol Ther* 82 (6): 756–759.

Ramagopalan, S. V., et al. (2009). "Expression of the Multiple Sclerosis-Associated MHC Class II Allele HLA-DRB1*1501 Is Regulated by Vitamin D." *PLoS Genet* 5 (2): e1000369.

Mansbach, J. M., et al. (2009). "Serum 25-Hydroxyvitamin D Levels Among US Children Aged 1 to 11 Years: Do Children Need More Vitamin D?" *Pediatrics* 124 (5): 1404–1410.

Caspi, A., et al. (2002). "Role of Genotype in the Cycle of Violence in Maltreated Children." *Science* 297 (5582): 851–854.

Meyer-Lindenberg, A., et al. (2006). "Neural Mechanisms of Genetic Risk for Impulsivity and Violence in Humans." *Proc Nat Acad Sci USA* 103 (16): 6269–6274.

Beaver, K. M., et al. (2009). "Monoamine Oxidase A Genotype Is Associated with Gang Membership and Weapon Use." *Compr Psychiatry* doi: 10.1016/j.comppsych.2009.03.010.

Kim-Cohen, J., et al. (2006). "MAOA, Maltreatment, and Gene-Environment Interactions Predicting Children's Mental Health: New Evidence and a Meta-Analysis." *Mol Psychiatry* 11: 903–913.

Kent, W. N., et al. (2006). "Role of Monoamine Oxidase A Genotype and Psychosocial Factors in Male Adolescent Criminal Activity." *Biol Psychiatry* 59 (2): 121–127.

Harold, D., et al. (2006). "Further Evidence That the KIAA0319 Gene Confers Susceptibility to Developmental Dyslexia." *Mol Psychiatry* 11: 1085–1091.

Paracchini, S., et al. (2008). "Association of the KIAA0319 Dyslexia Susceptibility Gene with Reading Skills in the General Population." *Am J Psychiatry* 165 (12): 1576–1584.

Gizer, I. R., et al. (2009). "Candidate Gene Studies of ADHD: A Meta-Analytic Review." *Hum Genet* 126 (1): 51–90.

Nathoo, N., et al. (2003). "Genetic Vulnerability Following Traumatic Brain Injury: The Role of Apolipoprotein E." *Mol Pathol* 56 (3): 132–136.

Lo, T., et al. (2009). "Modulating Effect of Apolipoprotein E Polymorphisms on Secondary Brain Insult and Outcome After Childhood Brain Trauma." *Childs Nerv Syst* 25 (1): 47–54.

National Football League. "NFL Player Care," 11/15/09: www.nflplayercare.com.

Schwarz, A. "Wives United by Husbands' Post-NFL Trauma." *New York Times*, 3/14/07.

"35 NFL Players Qualify for Dementia-Alzheimer's Assistance." FoxNews.com, 5/31/07: www.foxnews.com/story/0,2933,276617,00.html.

Weir, D. (2009). "Retired NFL Player's Survey." The University of Michigan's Institute for Social Research, Michigan Center on the Demography of Aging.

Schwarz, A. "NFL Study Finds Link to Dementia." *New York Times*, 9/29/09.

Strobel, G., et al. "Live Discussion: Sports Concussions, Dementia, and APOE Genotyping: What Can Scientists Tell the Public? What's Up for Research?" Alzheimer Research Forum, 11/11/08.

Schwarz, A. "Silence on Concussions Raises Risks of Injury." *New York Times*, 9/15/07.

Van Eerdewegh, P., et al. (2002). "Association of the ADAM33 Gene with Asthma and Bronchial Hyperresponsiveness." *Nature* 418 (6896): 426–430.

Reijmerink, N. E., et al. (2009). "Smoke Exposure Interacts with ADAM33 Polymorphisms in the Development of Lung Function and Hyperresponsiveness." *Allergy* 64 (6): 898–904.

Tantisira, K. G., et al. (2004). "Corticosteroid Pharmacogenetics: Association of Sequence Variants in CRHR1 with Improved Lung Function in Asthmatics Treated with Inhaled Corticosteroids." *Hum Mol Genet* 13 (13): 1353–1359.

Fligor, B. J., et al. (2004). "Output Levels of Commercially Available Portable Compact Disc Players and the Potential Risk to Hearing." *Ear Hear* 25 (6): 513–527.

Taibbi, M. "Can Your iPod Cause You to Go Deaf? A Doctor Hopes His Advice on Limiting Music Volume Doesn't Fall on Deaf Ears." NBC News, 8/3/05.

Konings, A., et al. (2008). "Variations in HSP70 Genes Associated with Noise-Induced Hearing Loss in Two Independent Populations." *Eur J Hum Genet* 17 (3): 329–335.

Preston, D. L., et al. (2002). "Radiation Effects on Breast Cancer Risk: A Pooled Analysis of Eight Cohorts." *Radiation Research* 158 (2): 220–235.

Berrington de Gonzalez, A., et al. (2009). "Estimated Risk of Radiation-Induced Breast Cancer from Mammographic Screening for Young BRCA Mutation Carriers." *J Natl Cancer Inst* 101 (3): 205–209.

Andrieu, N., et al. (2006). "Effect of Chest X-rays on the Risk of Breast Cancer Among BRCA1/2 Mutation Carriers in the International BRCA1/2 Carrier Cohort Study: A Report from the EMBRACE, GENEPSO, GEO-HEBON, and IBCCS Collaborators' Group." *J Clin Oncol* 24 (21): 3361–3366.

Esther, M. J., et al. (2007). "Medical Radiation Exposure and Breast Cancer Risk: Findings from the Breast Cancer Family Registry." *Int J Cancer* 121 (2): 386–394.

Parveen, B., et al. (2008). "Polymorphisms in DNA Repair Genes, Ionizing Radiation Exposure and Risk of Breast Cancer in U.S. Radiologic Technologists." *Int J Cancer* 122 (1): 177–182.

Lessov-Schlaggar, C. N., et al. (2008). "Genetics of Nicotine Dependence and Pharmacotherapy." *Biochem Pharmacol* 75 (1): 178–195.

8. Protecting Your Cardiovascular Health by Beating Your DNA

Renaud, S., et al. (1992). "Wine, Alcohol, Platelets, and the French Paradox for Coronary Heart Disease." *Lancet* 339 (8808): 1523–1526.

Hofman, N., et al. (2007). "Contribution of Inherited Heart Disease to Sudden Cardiac Death in Childhood." *Pediatrics* 120 (4): e967–e973.

Rodríguez-Calvo, M. S., et al. (2008). "Molecular Genetics of Sudden Cardiac Death." *Forensic Sci Int* 182 (1–3): 1–12.

Bos, J. M., et al. (2009). "Diagnostic, Prognostic, and Therapeutic Implications of Genetic Testing for Hypertrophic Cardiomyopathy." *J Am Coll Cardiol* 54 (3): 201–211.

Splawski, I., et al. (2002). "Variant of SCN5A Sodium Channel Implicated in Risk of Cardiac Arrhythmia." *Science* 297 (5585): 1333–1336.

Burke, A., et al. (2005). "Role of SCN5A Y1102 Polymorphism in Sudden Cardiac Death in Blacks." *Circulation* 112: 798–802.

Chen, S., et al. (2002). "SNP S1103Y in the Cardiac Sodium Channel Gene SCN5A Is Associated with Cardiac Arrhythmias and Sudden Death in a White Family." *J Med Genet* 39 (12): 913–915.

Plant, L. D. (2006). "A Common Cardiac Sodium Channel Variant Associated with Sudden Infant Death in African Americans, SCN5A S1103Y." *J Clin Invest* 116 (2): 430–435.

Helgadottir, A., et al. (2007). "A Common Variant on Chromosome 9p21 Affects the Risk of Myocardial Infarction." *Science* 316 (5830): 1491–1493.

Schaefer, A. S., et al. (2009). "Identification of a Shared Genetic Susceptibility Locus for Coronary Heart Disease and Periodontitis." *PLoS Genet* 5 (2): e1000378.

Emmerich, J., et al. (2001). "Combined Effect of Factor V Leiden and Prothrombin 20210A on the Risk of Venous Thromboembolism: Pooled Analysis of 8 Case-Control Studies Including 2310 Cases and 3204 Controls: Study Group for Pooled-Analysis in Venous Thromboembolism." *J Thromb Haemost* 86 (3): 809–816.

Arnett, D. K., et al. (2007). "Relevance of Genetics and Genomics for Prevention and Treatment of Cardiovascular Disease: A Scientific Statement from the American Heart Association Council on Epidemiology and Prevention, the Stroke Council, and the Functional Genomics and Translational Biology Interdisciplinary Working Group." *Circulation* 115 (22): 2878–2901.

Ye, Z., et al. (2006). "Seven Haemostatic Gene Polymorphisms in Coronary Disease: Meta-Analysis of 66,155 Cases and 91,307 Controls." *Lancet* 367 (9511): 651–658.

Elisabeth, R. P., et al. (2007). "Risk of Venous Thrombosis: Obesity and Its Joint Effect with Oral Contraceptive Use and Prothrombotic Mutations." *Br J Haematol* 139 (2): 289–296.

Spannagl, M. (2000). "Are Factor V Leiden Carriers who Use Oral Contraceptives at Extreme Risk for Venous Thromboembolism?" *Eur J Contracept Reprod Health Care* 5 (2): 105–112.

Bloom, M. "My Husband Should Be Living Today." MSNBC Today, 3/3/05.

Gschwendtner, A., et al. (2009). "Sequence Variants on Chromosome 9p21.3 Confer Risk of Atherosclerotic Stroke." *Ann Neurol* 65 (5): 531–539.

Gretarsdottir, S., et al. (2008). "Risk Variants for Atrial Fibrillation on Chromosome 4q25 Associate with Ischemic Stroke." *Ann Neurol* 64 (4): 402–409.

Bilguvar, K., et al. (2008). "Susceptibility Loci for Intracranial Aneurysm in European and Japanese Populations." *Nat Genet* 40 (12): 1472–1477.

Small, K. M., et al. (2002). "Synergistic Polymorphisms of $ß_1$- and α_{2C}-Adrenergic Receptors and the Risk of Congestive Heart Failure." *N Engl J Med* 347 (15): 1135–1142.

Lobmeyer, M. T., et al. (2007). "Synergistic Polymorphisms of Beta-1 and Alpha-2C-Adrenergic Receptors and the Influence on Left Ventricular Ejection Fraction Response to Beta-Blocker Therapy in Heart Failure." *Pharmacogenet Genomics* 17 (4): 277–282.

Lorenz, E. N. (1993). "Predictability: Does the Flap of a Butterfly's Wings in Brazil Set Off a Tornado in Texas?" *The Essence of Chaos* (The Jessie and John Danz Lecture Series) (Seattle: University of Washington Press), 181–184.

Dhandapany, P. S., et al. (2009). "A Common MYBPC3 (Cardiac Myosin Binding Protein C) Variant Associated with Cardiomyopathies in South Asia." *Nat Genet* 41 (2): 187–191.

Svetkey, L. P., et al. (2001). "Angiotensinogen Genotype and Blood Pressure Response in the Dietary Approaches to Stop Hypertension (DASH) Study." *J Hypertens* 19: 1949–1956.

Hassan, M., et al. (2008). "Association of Beta-1-Adrenergic Receptor Genetic Polymorphism with Mental Stress-Induced Myocardial Ischemia in Patients with Coronary Artery Disease." *Arch Intern Med* 168 (7): 763–770.

Warner, J. H. (1986). *The Therapeutic Perspective: Medical Practice, Knowledge and Identity in America, 1828–1885* (Cambridge: Harvard University Press), 28, 33.

Ingelman-Sundberg, M. (2001). "Pharmacogenetics: An Opportunity for a Safer and More Efficient Pharmacotherapy." *J Intern Med* 250 (3): 186–200.

Georgirene, D. V., et al. (2006). "Genetic Risk Factors Associated with Lipid-Lowering Drug-Induced Myopathies." *Muscle Nerve* 34 (2): 153–162.

Chasman, D. I., et al. (2004). "Pharmacogenetic Study of Statin Therapy and Cholesterol Reduction." *JAMA* 291 (23): 2821–2827.

Nakamura, Y. (2008). "Pharmacogenomics and Drug Toxicity." *N Engl J Med* 359 (8): 856–858.

Shuldiner, A. R., et al. (2009). "Association of Cytochrome P450 2C19 Genotype with the Antiplatelet Effect and Clinical Efficacy of Clopidogrel Therapy." *JAMA* 302 (8): 849–857.

DeGeorge, B. R., et al. (2007). "Beta-Blocker Specificity: A Building Block Toward Personalized Medicine." *J Clin Invest* 117 (1): 86–89.

Cooke, G. E., et al. (1998). "Pl^A2 Polymorphism and Efficacy of Aspirin." *Lancet* 351 (9111): 1253.

Caraco, Y., et al. (2008). "CYP2C9 Genotype-Guided Warfarin Prescribing Enhances the Efficacy and Safety of Anticoagulation: A Prospective Randomized Controlled Study." *Clin Pharmacol Ther* 83: 460–470.

U.S. Food and Drug Administration. "Critical Path Initiative—Warfarin Dosing," 10/13/09: www.fda.gov/ScienceResearch/SpecialTopics/CriticalPathInitiative/FacesBehindCriticalPath/ucm077473.htm.

9. A New Strategy in Our War Against Alzheimer's Disease

Avramopoulos, D. (2009). "Genetics of Alzheimer's Disease: Recent Advances." *Genome Med* 1 (3): 34.

Alzheimer, A. (1907). "Ueber eine eigenartige Erkrankung der Himrinde." *Allg Z Psychiat Med* 64: 146–148.

Jacob, R., et al. (2004). "ApoE Genotype Accounts for the Vast Majority of AD Risk and AD Pathology." *Neurobiol Aging* 25 (5): 641–650.

Caselli, R. J., et al. (2007). "Cognitive Domain Decline in Healthy Apolipoprotein E ε4 Homozygotes Before the Diagnosis of Mild Cognitive Impairment." *Arch Neurol* 64: 1306–1311.

Li, Y., et al. (2008). "SORL1 Variants and Risk of Late-Onset Alzheimer's Disease." *Neurobiol Dis* 29: 293–296.

Ikram, M. A., et al. (2009). "The GAB2 Gene and the Risk of Alzheimer's Disease: Replication and Meta-Analysis." *Biol Psychiatry* 65 (11): 995–999.

Bertram, L., et al. (2008). "Genome-Wide Association Analysis Reveals Putative Alzheimer's Disease Susceptibility Loci in Addition to APOE." *Am J Hum Genet* 83: 623–632.

Roses, A. D., et al. (2009). "Apoe-3 and Tomm-40 Haplotypes Determine Inheritance of Alzheimer's Disease Independently of Apoe-4 Risk." Alzheimer's Association 2009 International Conference on Alzheimer's Disease, Vienna Austria, presentation O1-06-01 on 7/12/09, 3–5 p.m.

Sigrid Botne, S., et al. (2008). "Risk-Reducing Effect of Education in Alzheimer's Disease." *Int J Geriatr Psychiatry* 23 (11): 1156–1162.

Valenzuela, M. J., et al. (2008). "Brain Reserve and the Prevention of Dementia." *Curr Opin Psychiatry* 21 (3): 296–302.

Friedman, G., et al. (1999). "Apolipoprotein E-epsilon4 Genotype Predicts a Poor Outcome in Survivors of Traumatic Brain Injury." *Neurology* 52 (2): 244–248.

Wang, H., et al. (2007). "An Apolipoprotein E-based Therapeutic Improves Outcome and Reduces Alzheimer's Disease Pathology Following Closed Head Injury: Evidence of Pharmacogenomic Interaction." *Neuroscience* 144 (4): 1324–1333.

Deeny, S. P., et al. (2008). "Exercise, APOE, and Working Memory: MEG and Behavioral Evidence for Benefit of Exercise in Epsilon4 Carriers." *Biol Psychol* 78 (2): 179–187.

Lautenschlager, N. T., et al. (2008). "Effect of Physical Activity on Cognitive Function in Older Adults at Risk for Alzheimer Disease: A Randomized Trial." *JAMA* 300 (9): 1027–1037.

Scarmeas, N., et al. (2006). "Mediterranean Diet and Risk for Alzheimer's Disease." *Ann Neurol* 59: 912–921.

Vingtdeux, V., et al. (2008). "Therapeutic Potential of Resveratrol in Alzheimer's Disease." *BMC Neuroscience* 9 (Suppl 2): S6.

Kim, D., et al. (2007). "SIRT1 Deacetylase Protects Against Neurodegeneration in Models for Alzheimer's Disease and Amyotrophic Lateral Sclerosis." *EMBO J* 26 (13): 3169–3179.

Rosso, A., et al. (2008). "Caffeine: Neuroprotective Functions in Cognition and Alzheimer's Disease." *Am J Alzheimers Dis Other Demen* 23 (5): 417–422.

Zandi, P. P., et al. (2002). "Reduced Incidence of AD with NSAID but Not H2 Receptor Antagonists: The Cache County Study." *Neurology* 59 (6): 880–886.

Li, G., et al. (2007). "Statin Therapy Is Associated with Reduced Neuropathologic Changes of Alzheimer Disease." *Neurology* 69 (9): 878–885.

Vuletic, S., et al. (2006). "Statins of Different Brain Penetrability Differentially Affect CSF PLTP Activity." *Dement Geriatr Cogn Disord* 22 (5–6): 392–398.

Dufouil, C., et al. (2005). "APOE Genotype, Cholesterol Level, Lipid-Lowering Treatment, and Dementia: The Three-City Study." *Neurology* 64 (9): 1531–1538.

Miller, E. R., et al. (2005). "Meta-Analysis: High-Dosage Vitamin E Supplementation May Increase All-Cause Mortality." *Ann Int Med* 142 (1): 37–46, and article correspondences.

"FDA Announces Qualified Health Claims for Omega-3 Fatty Acids." FDA News Release, 9/8/04.

Huang, T. L., et al. (2005). "Benefits of Fatty Fish on Dementia Risk Are Stronger for Those Without APOE Epsilon4." *Neurology* 65 (9): 1409–1414.

Bonnie, J. H. (2007). "New Studies Support the Therapeutic Value of Meditation." *Explore* (New York) 3 (5): 449–452.

Luders, E., et al. (2009). "The Underlying Anatomical Correlates of Long-Term Meditation: Larger Hippocampal and Frontal Volumes of Gray Matter." *NeuroImage* 45 (3): 672–678.

Peavy, G. M., et al. (2007). "The Effects of Prolonged Stress and APOE Genotype on Memory and Cortisol in Older Adults." *Biol Psychiatry* 62 (5): 472–478.

Pike, K. E., et al. (2007). "Beta-Amyloid Imaging and Memory in Nondemented Individuals: Evidence for Preclinical Alzheimer's Disease." *Brain* 130: 2837–2844.

Cacabelos, R. (2007). "Donepezil in Alzheimer's Disease: From Conventional Trials to Pharmacogenetics." *Neuropsychiatr Dis Treat* 3 (3): 303–333.

10. Predict, Prevent, and Prevail over Cancer

(1996). "The Edwin Smith Surgical Papyrus" (1600 BC): http://touregypt.net/edwinsmithsurgical.htm.

Cho, M. K., et al. (1999). "Commercialization of BRCA1/2 Testing: Practitioner Awareness and Use of a New Genetic Test." *Am J Med Genet* 83 (3): 157–163.

Aspinwall, L. G., et al. (2008). "CDKN2A/p16 Genetic Test Reporting Improves Early Detection Intentions and Practices in High-Risk Melanoma Families." *Cancer Epidemiol Biomarkers Prev* 17 (6): 1510–1519.

Kerzendorfer, C., et al. (2009). "UVB and Caffeine: Inhibiting the DNA Damage Response to Protect Against the Adverse Effects of UVB." *J Invest Dermatol* 129 (7): 1611–1613.

Kennedy, C., et al. (2001). "Melanocortin 1 Receptor (MC1R) Gene Variants Are Associated with an Increased Risk for Cutaneous Melanoma Which Is Largely Independent of Skin Type and Hair Color." *J Invest Dermatol* 117: 294–300.

Giudice, M. F. Personal communication, 8/10/09.

Binkley, C. J., et al. (2009). "Genetic Variations Associated with Red Hair Color and Fear of Dental Pain, Anxiety Regarding Dental Care and Avoidance of Dental Care." *J Am Dent Assoc* 140 (7): 896–905.

Ognjanovic, S., et al. (2006). "NAT2, Meat Consumption and Colorectal Cancer Incidence: An Ecological Study Among 27 Countries." *Cancer Causes Control* 17 (9): 1175–1182.

Samad, A. K., et al. (2005). "A Meta-Analysis of the Association of Physical Activity with Reduced Risk of Colorectal Cancer." *Colorectal Dis* 7 (3): 204–213.

Baron, J. A., et al. (2003). "A Randomized Trial of Aspirin to Prevent Colorectal Adenomas." *New Engl J Med* 348: 891–899.

Ngo, S. N. T., et al. (2007). "Does Garlic Reduce Risk of Colorectal Cancer? A Systematic Review." *J Nutr* 137 (10): 2264–2269.

Zheng, S. L., et al. (2008). "Cumulative Association of Five Genetic Variants with Prostate Cancer." *N Engl J Med* 358 (9): 910–919.

Xu, J., et al. (2009). "Estimation of Absolute Risk for Prostate Cancer Using Genetic Markers and Family History." *Prostate* 69 (14): 1565–1572.

Thompson, I. M., et al. (2003). "The Influence of Finasteride on the Development of Prostate Cancer." *N Engl J Med* 349 (3): 215–224.

Pohar, K. S., et al. (2003). "Tomatoes, Lycopene and Prostate Cancer: A Clinician's Guide for Counseling Those at Risk for Prostate Cancer." *World J Urol* 21 (1): 9–14.

Fradet, V., et al. (2009). "Dietary Omega-3 Fatty Acids, Cyclooxygenase-2 Genetic Variation, and Aggressive Prostate Cancer Risk." *Clin Cancer Res* 15 (7): 2559–2566.

Spitz, M. R., et al. (2003). "Genetic Susceptibility to Lung Cancer: The Role of DNA Damage and Repair." *Cancer Epidemiol Biomarkers Prev* 12: 689–698.

Lee, A. M., et al. (2007). "CYP2B6 Genotype Alters Abstinence Rates in a Bupropion Smoking Cessation Trial." *Biol Psychiatry* 62 (6): 635–641.

Lerman, C., et al. (2006). "Role of Functional Genetic Variation in the Dopamine D2 Receptor (DRD2) in Response to Bupropion and Nicotine Replacement Therapy for Tobacco Dependence: Results of Two Randomized Clinical Trials." *Neuropsychopharmacol* 31 (1): 231–242.

Johnstone, E. C., et al. (2004). "Genetic Variation in Dopaminergic Pathways and Short-Term Effectiveness of the Nicotine Patch." *Pharmacogenetics* 14: 83–90.

Brennan, P., et al. (2005). "Effect of Cruciferous Vegetables on Lung Cancer in Patients Stratified by Genetic Status: A Mendelian Randomisation Approach." *Lancet* 366 (9496): 1558–1560.

Batra, V., et al. (2003). "The Genetic Determinants of Smoking." *Chest* 123: 1730–1739.

Richard, S. W., et al. (2008). "Association of a Single Nucleotide Polymorphism in Neuronal Acetylcholine Receptor Subunit Alpha 5 (CHRNA5) with Smoking Status and with 'Pleasurable Buzz' During Early Experimentation with Smoking." *Addiction* 103 (9): 1544–1552.

Isabel, R. S., et al. (2008). "The CHRNA5/A3/B4 Gene Cluster Variability as an Important Determinant of Early Alcohol and Tobacco Initiation in Young Adults." *Biol Psychiatry* 63 (11): 1039–1046.

Han, S. W., et al. (2005). "Predictive and Prognostic Impact of Epidermal Growth Factor Receptor Mutation in Non-Small-Cell Lung Cancer Patients Treated with Gefitinib." *J Clin Oncol* 23: 2493–2501.

Antoniou, A., et al. (2003). "Average Risks of Breast and Ovarian Cancer Associated with BRCA1 or BRCA2 Mutations Detected in Case Series Unselected for Family History: A Combined Analysis of 22 Studies." *Am J Hum Genet* 72 (5): 1117–1130.

Bleyer, A., et al. (2006). "Cancer in 15- to 29-Year-Olds by Primary Site." *Oncologist* 11 (6): 590–601.

Meiser, B., et al. (2002). "Psychological Impact of Genetic Testing in Women from High-Risk Breast Cancer Families." *Eur J Cancer* 38: 2025–2031.

Angela, R. B., et al. (2009). "Learning of Your Parent's BRCA Mutation During Adolescence or Early Adulthood: A Study of Offspring Experiences." *Psycho-Oncology* 18 (2): 200–208.

Parthasarathy, S. (2005). "Architectures of Genetic Medicine: Comparing Genetic Testing for Breast Cancer in the USA and the UK." *Soc Stud Sci* 35 (1): 5–40.

John, E. M., et al. (2007). "Prevalence of Pathogenic BRCA1 Mutation Carriers in 5 US Racial/Ethnic Groups." *JAMA* 298 (24): 2869–2876.

Antoniou, A. C., et al. (2007). "RAD51 135G→C Modifies Breast Cancer Risk Among BRCA2 Mutation Carriers: Results from a Combined Analysis of 19 Studies." *Am J Hum Genet* 81 (6): 1186–1200.

CHEK2 Breast Cancer Case-Control Consortium. (2004). "CHEK2*1100delC and Susceptibility to Breast Cancer: A Collaborative Analysis Involving 10,860 Breast Cancer Cases and 9,065 Controls from 10 Studies." *Am J Hum Genet* 74 (6): 1175–1182.

Walsh, T., et al. (2006). "Spectrum of Mutations in BRCA1, BRCA2, CHEK2, and TP53 in Families at High Risk of Breast Cancer." *JAMA* 295 (12): 1379–1388.

Johnson, N., et al. (2007). "Counting Potentially Functional Variants in BRCA1, BRCA2 and ATM Predicts Breast Cancer Susceptibility." *Hum Mol Genet* 16 (9): 1051–1057.

Sellers, T. A., et al. (2001). "Dietary Folate Mitigates Alcohol Associated Risk of Breast Cancer in a Prospective Study of Postmenopausal Women." *Epidemiology* 12 (4): 420–428.

Zheng, T., et al. (2003). "Glutathione S-transferase M1 and T1 Genetic Polymorphisms, Alcohol Consumption and Breast Cancer Risk." *Br J Cancer* 88 (1): 58–62.

McTiernan, A., et al. (2003). "Women's Health Initiative Cohort Study. Recreational Physical Activity and the Risk of Breast Cancer in Postmenopausal Women: The Women's Health Initiative Cohort Study." *JAMA* 290 (10): 1331–1336.

Wu, A. H., et al. (2003). "Tea Intake, COMT Genotype, and Breast Cancer in Asian-American Women." *Cancer Res* 63 (21): 7526–7529.

Kotsopoulos, J., et al. (2007). "The CYP1A2 Genotype Modifies the Association Between Coffee Consumption and Breast Cancer Risk Among BRCA1 Mutation Carriers." *Cancer Epidemiol Biomarkers Prev* 16 (5): 912–916.

Jernström, H., et al. (2004). "Breast-Feeding and the Risk of Breast Cancer in BRCA1 and BRCA2 Mutation Carriers." *J Natl Cancer Inst* 96 (14): 1094–1098.

Cullinane, C. A., et al. (2005). "Effect of Pregnancy as a Risk Factor for Breast Cancer in BRCA1/BRCA2 Mutation Carriers." *Int J Cancer* 117 (6): 988–991.

Warner, E., et al. (2004). "Surveillance of BRCA1 and BRCA2 Mutation Carriers with Magnetic Resonance Imaging, Ultrasound, Mammography, and Clinical Breast Examination." *JAMA* 292 (11): 1317–1325.

Berrington de Gonzalez, A., et al. (2009). "Estimated Risk of Radiation-Induced Breast Cancer from Mammographic Screening for Young BRCA Mutation Carriers." *J Natl Cancer Inst* 101 (3): 205–209.

Andrieu, N., et al. (2006). "Effect of Chest X-rays on the Risk of Breast Cancer Among BRCA1/2 Mutation Carriers in the International BRCA1/2 Carrier Cohort Study: A Report from the EMBRACE, GENEPSO, GEO-HEBON, and IBCCS Collaborators' Group." *J Clin Oncol* 24 (21): 3361–3366.

Saslow, D., et al. (2007). "American Cancer Society Guidelines for Breast Screening with MRI as an Adjunct to Mammography." *CA Cancer J Clin* 57 (2): 75–89.

Hayes, D. F., et al. (2008). "Consortium on Breast Cancer Pharmacogenomics. A Model Citizen? Is Tamoxifen More Effective Than Aromatase Inhibitors If We Pick the Right Patients?" *J Natl Cancer Inst* 100 (9): 610–613.

Tan, S.-C., et al. (2008). "Pharmacogenetics in Breast Cancer Therapy." *Clin Cancer Res* 14: 8027–8041.

Schroth, W., et al. (2007). "Breast Cancer Treatment Outcome with Adjuvant Tamoxifen Relative to Patient CYP2D6 and CYP2C19 Genotypes." *J Clin Oncol* 25 (33): 5187–5193.

National Society of Genetic Counselors. NSGC Listserv discussion, 7/29/09.

Metcalfe, K. A., et al. (2005). "The Use of Preventive Measures Among Healthy Women who Carry a BRCA1 or BRCA2 Mutation." *Fam Cancer* 4 (2): 97–103.

Evans, D. G. R., et al. (2009). "Risk Reducing Mastectomy: Outcomes in 10 European Centres." *J Med Genet* 46 (4): 254–258.

Frost, M. H., et al. (2000). "Long-Term Satisfaction and Psychological and Social Function Following Bilateral Prophylactic Mastectomy." *JAMA* 284 (3): 319–324.

Kramer, J. L., et al. (2005). "Prophylactic Oophorectomy Reduces Breast Cancer Penetrance During Prospective, Long-Term Follow-Up of BRCA1 Mutation Carriers." *J Clin Oncol* 23 (34): 8629–8635.

Afterword

Brazell, C., et al. (2002). "Maximizing the Value of Medicines by Including Pharmacogenetic Research in Drug Development and Surveillance." *Br J Clin Pharmacol* 53 (3): 224–231.

Lesko, L. J., et al. (2003). "Pharmacogenetics and Pharmacogenomics in Drug Development and Regulatory Decision Making: Report of the First FDA-PWG-PhRMA-DruSafe Workshop." *J Clin Pharmacol* 43 (4): 342–358.

Roses, A. D. (2002). "Genome-Based Pharmacogenetics and the Pharmaceutical Industry." *Nat Rev Drug Discov* 1 (7): 541–549.

Reeves, I., et al. (2006). "Screening for HLA-B*5701 Reduces the Frequency of Abacavir Hypersensitivity." *Antivir Ther* 11: L11.

U.S. Food and Drug Administration. "Information for Healthcare Professionals: Abacavir (Marketed as Ziagen) and Abacavir-Containing Medications." FDA Alert, 7/24/08.

Townes, T. M. (2008). "Gene Replacement Therapy for Sickle Cell Disease and Other Blood Disorders." *Hematology* 2008 (1): 193–196.

Cideciyan, A. V., et al. (2009). "Vision 1 Year After Gene Therapy for Leber's Congenital Amaurosis." *N Engl J Med* 361 (7): 725–727.

Miller Jr., B. L. Personal communication, 9/3/09.

Behar, D. M., et al. (2008). "The Dawn of Human Matrilineal Diversity." *Am J Hum Genet* 82 (5): 1130–1140.

Kurzweil, R. (2007). "The Singularity Is Near: When Humans Transcend Biology." Penguin Kindle ed.

Keehan, S. et al. (2008). "Health Spending Projections Through 2017." *Health Aff* 27 (2): w145–w155.

Pear, R. "U.S. Health Care Spending Reaches All-Time High: 15% of GDP." *New York Times*, 1/9/04.

U.S. Centers for Disease Control and Prevention. "Costs of Chronic Disease." *Chronic Disease Overview*, 11/14/09: www.cdc.gov/nccdphp/overview.htm.

Langheier, J. M., et al. (2004). "Prospective Medicine: The Role for Genomics in Personalized Health Planning." *Pharmacogenomics* 5 (1): 1–8.

Hood, L. (2009). "Medicine of the Future: The Transformation from Reactive to Proactive (P4) Medicine: Predictive, Personalized, Preventive, and Participatory." Institute for Systems Biology, Seattle, WA.

Index

Page numbers in **bold** indicate tables; those in *italics* indicate figures.